T0314249

Green Capital

Green Capital

A New Perspective on Growth

Christian de Perthuis
and Pierre-André Jouvet

Translated by Michael Westlake

COLUMBIA UNIVERSITY PRESS

NEW YORK

Columbia University Press
Publishers Since 1893
New York   Chichester, West Sussex
Copyright © 2013 Odile Jacob

Translation copyright © 2015 Columbia University Press
All rights reserved

Library of Congress Cataloging-in-Publication Data
Perthuis, Christian de.
[Capital vert. English]
Green capital : a new perspective on growth / Christian de Perthuis,
Pierre-André Jouvet; translated by Michael Westlake.
pages cm
Translation of: Le capital vert : une nouvelle perspective
de croissance, published in 2013.
Includes bibliographical references and index.
ISBN 978-0-231-17140-3 (cloth : alk. paper)
ISBN 978-0-231-54036-0 (e-book)
1. Environmental economics. 2. Sustainable development.
I. Jouvet, Pierre-André. II. Title.

HC79.E5P45713 2015
333.7—dc23
2015002297

Columbia University Press books are printed on permanent
and durable acid-free paper.
This book is printed on paper with recycled content.
Printed in the United States of America

c 10 9 8 7 6 5 4 3 2 1

Cover design: Noah Arlow

Contents

Contents

Contents

Green Capital

The Color of Growth

FOUR THOUSAND YEARS AGO, the fate of the Sumerians revealed that when growth is driven by capital accumulation that preys on the environment, it will eventually self-destruct. Thanks to their control of irrigation, the inhabitants of Sumer developed agriculture, writing, law, and the city; but because they were unable to master drainage, the indispensable complement to irrigation in arid areas, their civilization vanished as a result of the sterilization of the soil by salt. The inhabitants of Easter Island, who had also developed one of the first writing systems, experienced a similar fate.[1] After felling the last tree, they abandoned their island with its huge granite statues and its now lunar landscape. In a mineral world, only stone can survive. Yet in cutting down the trees, the islanders not only exhausted a supply of resources, but, as Jared Diamond so brilliantly shows, they also destroyed the reproductive capacity of an ecosystem.[2]

As with the Sumerians and the Easter Islanders, the world's economic growth is at risk of faltering. For millennia, things like the water cycle, ecosystems diversity, and the greenhouse effect have helped shape the conditions of life on Earth by freely providing what is required for the reproduction of resources. These can be thought of as "regulatory systems" that help protect and enhance our "natural capital," that is, the stock of natural ecosystems that generates a flow of valuable goods or services. As they function today, markets are destroying these regulatory systems, and in recent decades there has been growing awareness of such threats. In 2006, in a widely publicized report, the economist Nicolas Stern estimated the cost of potential losses to the world economy by 2050 as a result of climate change at up to 20 percent of gross domestic product (GDP). Some years later, the Sukhdev report (prepared for the Nagoya conference of 2011) estimated that the services provided by the diversity of ecosystems amounted to 40 percent of global GDP.[3] Scientific reports all reveal the rapid loss of this biodiversity. Suppose that the services currently provided for free by biodiversity were reduced by a quarter between now and 2050; this would amount to a 10 percent decrease in global GDP, probably irreversibly.

Despite the supportive discourse of international organizations like the OECD and the World Bank, which has lent credibility to the idea of "green growth," these new environmental concerns remain on the periphery of political and economic decision making. Worse, following the deep recession of 2008–2009, the outlook of decision makers has shortened: what counts now is a rapid return to growth and the reduction of unemployment. As for the color of growth, they seem to say, we'll think about that later! Barack Obama's first presidential campaign in the fall of 2008 focused on two societal projects: the extension of health coverage and controlling greenhouse gas emissions. His 2012 campaign revolved around one point only: who, given the choice between the incumbent president and his Republican opponent, would be most able to stimulate the economy and create jobs? A few months earlier, the environment had likewise disappeared from the debate around the French presidential campaign. International life is in step with this restricting of the field of vision. In 2009, when the Copenhagen summit brought together many heads of state, climate change still seemed to be a major challenge for policy makers. The economic and financial crisis has since taken its toll. By 2013, the marathon sessions that counted were those that sought to save the euro or to defend the quality of one's sovereign debt—in short, to repair the economic machine and quickly restore growth. Climate change, ecology, the color of growth: these are no longer on the agenda.

In adopting such an approach, the industrialized world is indefinitely postponing any ambitious action to address climate change and the challenges presented by the environment more generally. With varying degrees of awareness, we are passing on the perils of global warming and loss of biodiversity to future generations. But by buttressing ourselves with such narrow visions of a return to growth, we are at the same time depriving ourselves of the most appropriate way of emerging from the current economic depression. Everyone vaguely feels that emergence from the crisis will not happen by copying past formulas. This time, exit requires a new wave of investment and innovation that will reconfigure the economic structure, as has taken place historically with the advent of animal traction, the steam engine, electricity, the computer, and the Internet. It is our personal conviction that "green capital" will play a central role in the reconstruction of the economy. But for this to happen, it is essential to stop pushing it to the periphery, and instead to place it at the heart of economic debates. The aim of this book is to contribute to this process by providing the reader with an itinerary consisting of five main stages.

The first four chapters seek to refine the diagnosis of complex relationships linking natural capital to economic growth. Some forty years ago, the celebrated Club of Rome report, entitled *The Limits to Growth*,[4] drew attention to the physical limits to growth imposed by the finiteness of natural resources. It is evident that human ingenuity and the capacity for innovation have disproven the report's predictions. The pace of global growth has scarcely diminished since the end of the post-war boom;[5] its center of gravity has simply shifted to China and newly emerging countries. Nor has growth been held back by any shortage of raw materials. Yet at the same time the impact of humanity on the natural environment has increased, threatening the major regulatory functions of natural capital such as climate stability and biodiversity. But these regulatory functions are incorporated into neither prices that calibrate values nor aggregates, such as GDP, that measure wealth.

Chapters 5 through 9 show how it is possible to move from a quantitative notion of the limits to growth based on the scarcity of natural resources to a panoptic outlook concerned with the preservation of the regulatory systems of natural capital. Rather than a representation deeply rooted in an economic concept of natural capital as a stock of scarce resources, there is a shift to a systemic view of natural capital, understood as a complex system of regulatory functions. Because nature is not a commodity that can be traded in a market, it is not possible to assign it a value, as is done for other components of capital. On the other hand, the deterioration of natural systems of regulation has a cost that reflects the use made of natural capital. We therefore include the cost of pollution in the production function because it contributes in the short term to the supply potential of the economy, although simultaneously weakening its long-term growth path. In the short term, pricing pollution changes the preexisting combination of production factors by attributing to the use of natural capital part of the supply previously attributed to labor and capital. By pricing pollution, green capital thus affects the short-term equilibrium and becomes a factor of production in which it is necessary to invest for long-term expansion. There are many ways of doing this. Adding a factor to the production function gives rise to an apportioning problem: Should it be capital or labor, the blue-collar worker or the boss, high-income countries or low-income countries who are obliged to cut back their revenue to pay for this new component of the natural capital that is progressively destroyed if a value, and hence a price, is not attached to it? This approach is consistent with the theoretical extensions that have progressively enriched the standard growth model developed in the 1950s by Robert Solow.

Chapters 10 through 13 present the assessment methods available for moving from the previously constructed growth model to an understanding of the concrete conditions for the transition to a green economy. This transition is still only in its infancy, with the first moves to introduce the value of natural capital into the economy now being taken. With regard to the climate system, the value collectively attributed to its preservation is measured by the costs associated with greenhouse gas emissions, more commonly termed the "carbon price." The methods for introducing this price into the economic system are now well known, but both nationally and internationally their implementation comes up against the need to manage the associated distributive affects. With regard to biodiversity, developing appropriate methods for pricing its uses is all the more complex because there is no equivalent to the "CO_2 standard" used for the climate. Various decentralized innovations and new economic research models will be needed to gradually incorporate into the economy the values that people want to attach to biodiversity. Without claiming to be exhaustive, this book attempts to identify, following pioneers like the British economist David Pearce,[6] the groundbreaking experiments that are opening up new fields for investment and thus growth.

Chapters 14 through 17 look in greater depth at the energy aspects of the transition to a green economy. First, various energy transition models are discussed: the U.S. energy revolution driven by the large-scale exploitation of oil and gas shale is a very different option than the low-carbon strategy adopted by Europe or the long-term diversification aimed at by oil-producing countries and the major emerging economies. The catalyzing role of energy pricing, and its climatic and environmental impacts, is then examined, along with the specific issue of nuclear power: though a non-CO_2-emitting primary energy source, it nonetheless gives rise to many other questions, as is shown by the French example. There follows an analysis of the various innovations that will accompany the energy transition: innovations in technology, of course, but also innovations in social organization, land management, and the way people live, along with innovations in terms of governance and the conduct of public affairs, without which public debates run the risk of being all talk—not having any real impact on policy decisions.

The final two chapters return to the concrete conditions required to foster growth based on the ascription of value to green capital, leading eventually to a self-reproducing, fully functional economy. The forces to be set in motion will emerge neither from spontaneous market action nor from deliberate action by planners. Instead their guiding principles will be the large-scale deployment of environmental pricing, reorientation of public support toward research and development, new choices in terms of infrastructure, and the introduction of greater intelligence into networks, as well as training, the organization of professional retraining, and social acceptance, without which a collective transition project cannot be constructed. These guiding principles are compared to the strategy adopted by Europe, followed by an exploration of the ways in which the European Union could become a real crucible for the ecological transition.

By way of conclusion to this itinerary, we investigate what type of radical shifts will result in the integration of green capital into the economy. Present-day economies are comparable to the situation of a shepherd on a mountainside shearing sheep on behalf of their owner. Making use of water from a stream running through the pasture, the shepherd produces a regular supply of fleece, thus enabling the owner to get a return on his capital and to pay the shepherd a wage. Capital and labor have been remunerated. But suppose that because of pollution, the water from the stream can no longer be used and the shepherd's productivity falls by half. This implies that natural capital was contributing free of charge to half of what was produced. The problem that arises is then very simple: who will pay for the shortfall corresponding to the loss of production? The shepherd, in which case, by extension, the wage rate in the economy declines? Or perhaps the owner, in which case the rate of profit falls? Toward which new economic paradigm does this book lead? Green capital or green capitalism?

Growth

A Historical Accident?

IN 1972, THE PUBLICATION of *The Limits to Growth* had a major impact. Commissioned by the Club of Rome, the book emerged from the work of a multidisciplinary modeling team from MIT directed by Dennis Meadows.[1] It showed that a number of exponential curves reflecting population growth, industrial production, the extraction of natural resources, etc. could not be extrapolated in a finite world. The growth process was therefore nothing more than a historical parenthesis and would come up against insurmountable limits. Its authors recommended anticipating the shock rather than running headlong into the wall of physical scarcity.

A year after the publication of the Meadows report, commodity prices were all soaring. The OPEC oil embargo led to fears of an overall shortage. The system of currency stability inherited from the Bretton Woods agreements,[2] signed after World War II, exploded, and the industrialized world experienced its first major recession since the war. Some people viewed this situation as evidence of the accuracy of Meadows's theses and as heralding the end of growth. They failed to foresee the new wave of expansion that would give rise to globalization, digital technologies, and the growing power of emerging countries. This expansionary wave lasted three decades before breaking with the recession of 2009.

Like that of 1973, the 2009 recession was the result of a twofold shock that wrecked the economic machine: commodity prices and financial deregulation.[3] Symbolically, the price of oil peaked in July 2008, two months before the bankruptcy of Lehman Brothers investment bank, an event that revealed to the world the extent of contamination of the financial system by excessive debt. Unlike with the Mexican and Asian crises in the preceding decades, it was no longer the periphery that was affected but the very heart of the system: Wall Street, with its mountain of subprime mortgages,[4] debt collateralized on an indefinite increase in price of property. Would this crisis come to mark, forty years after the publication of the Meadows report, the end of the period of rapid growth that began at the end of World War II? Would that "golden age," as it had been termed by the economist Angus Maddison, turn out to have been a mere accident of history?

Economic Growth Since 1500: Gradual Then Faster Acceleration

Considered solely from the quantitative standpoint, growth can be represented as the increase in a global magnitude measuring wealth. Take gross domestic product (GDP) as an indicator of this wealth, and transpose it to a per capita basis. In the long run, this indicator is very much a matter of convention because it involves comparing today's goods and services, many of which did not exist in the past, to yesterday's, which in many cases no longer exist in the present. The very notion of wealth, and its relationship to welfare, has to be treated with caution (a topic discussed in more detail in chapter 4). The observations made by the economists Maddison and Bairoch[5] from this type of indicator nevertheless provide valuable lessons.

According to Maddison and Bairoch, the world economy moved from a stationary regime to a growth regime in the fifteenth century, a time of major discoveries. Technological advances, mainly in agriculture and navigation, then led to a slow rise in productivity that in turn underpinned this first secular growth cycle.

Until around 1820, growth was barely detectable in the passage from one generation to the next: it took more than two hundred years for GDP per capita to double. The perception people would have had of their standard of living was that of a stationary regime: only the vagaries of climate, wars, and epidemics might make their lives end in better or worse conditions than at the time of their birth. The idea of progress was forged gradually with the thinkers of the Enlightenment, who began to foresee the potential impact of the application of technology on the organization of social life. But this idea remained visionary and uncertain and was held only by a minority. In the famed correspondence between Voltaire and Rousseau in the aftermath of the Lisbon earthquake of 1755, Rousseau expressed the prevailing view of the time that nature alone was responsible for such disasters, against which human ingenuity was powerless. It was not until the nineteenth century that the idea of progress became generally established with positivism and the birth of political economy.[6]

Another feature of this preindustrial period is that differences in living standards between countries and continents were nothing like what they became later. For example, it is difficult to assess whether the wealth per capita was higher in China or in Western Europe in the early nineteenth century.[7] But international trade in goods and capital was still very limited in the preindustrial world. It was not able to provide the function of equalizing living standards that generations of liberal economists would later attribute to it.

The first acceleration of growth began around 1820 in England and was rapidly transmitted to Western Europe and the United States.

With the taming and then the widespread use of new forms of energy—coal, then oil and electricity—the global economy moved up a gear. Product per capita grew by just under 1 percent per year, or a doubling every seventy years. This shift was reflected fairly quickly in living conditions and led to a decrease in mortality that in turn resulted in an increase in population growth. If we combine the wealth effect per capita with the increase in population, we obtain overall growth of around 1.75 percent per year: a doubling every forty years for a century and a half. Such a pace transforms living conditions, especially through urbanization. It was accompanied by a significant widening of the disparity between economic areas. Wealth became polarized as it never had been historically in Europe, North America, and Japan, which together represented about a fifth of the world population. At the same time, huge economic territories were destructured by European and American expansion. In particular, the magnitude of the Chinese and Indian economies was eroded during this period that coincided, moreover, with colonial exploitation.

The period following World War II saw a further—spectacular and unexpected—acceleration in growth. In the old industrialized countries (Europe, the United States, and Japan), the momentum was initiated by expenditure on making good the damage caused by the war. In addition, Western Europe and even more so Japan were among the economies benefiting most from this "golden age." What was new was that economic and population growth were diffused much more widely throughout the world. Overall, GDP growth per capita approached 3 percent per year, or a doubling every quarter century: a rate unprecedented in human history. Many people thought that this historical anomaly would end in 1970 with the breakdown of the Bretton Woods monetary system and the subsequent sharp recession as a result of the oil shock. History, however, followed another scenario: not that of a slowdown in global growth but of its redistribution among geographical regions.

Since 1973: Geographical Redistribution
of Global Growth

Forty years after the publication of *The Limits to Growth*, the weakening of growth in the old industrialized countries is not in doubt. Between 1973 and 2013, the slowdown in their trajectory was particularly marked in the case of Japan and more moderate in Europe and the United States. The decline reflects the difficulties of adjusting in a globalized world in which taking on debt is facilitated by deregulated capital markets and in which the sanctioning of their excesses may come out of the blue because it is no longer possible to eliminate debt mountains through the traditional means of inflation. Japan was the first to fall into this deflationary trap in the 1990s. It was followed by the English-speaking countries, where the great recession of 2008–2009 originated, with the latest episode being the euro sovereign debt crisis from 2011 onward.

The slowdown in these economies has not, however, resulted in weaker global growth because it has been offset by the economic dynamism of emerging countries. By accumulating capital at an unprecedented pace, China, India, Brazil, and some other countries have, since 1973, been gradually making up for their secular backwardness. New forces are in operation and account for the speed of such redistribution. The openness of trade has acquired a new dimension. In 1950, the proportionate volume of goods traded in the global economy was below its 1913 level; over the next fifty years it increased fourfold. Following the liberalization of markets, financial capital has become totally mobile. Information technologies facilitate the acquisition of new expertise, the spread of innovations, and the interconnectedness of markets. Beginning with the takeoff of small Asian countries, this epochal change extended to mainland China, which started to open up its markets in 1979 with the coming to power of Deng Xiaoping, followed by the other continent-sized country, India.

The propagation of this dynamic to other regions of the world did not take place without its ups and downs. It took Latin America a good decade to adjust to the two oil shocks, culminating in the Mexican debt crisis, which convulsed the continent during the 1980s. Africa was even worse affected, with a fall in GDP per capita between 1973 and 1998. The transition of the centralized economy of the USSR to a market economy took place under the worst of circumstances. The disintegration of the state and the application of free market recipes naively transposed from the University of Chicago caused an implosion of the Russian economy and those of its former satellites in the 1990s.

The first decade of the new millennium amplified the geographical spread of global growth, whose center of gravity shifted to China and emerging countries. Latin America emerged from its difficulties at the beginning of the 1990s. Russia and its former satellites managed to halt their downward spiral around 1997–1998. Since 2001, sub-Saharan Africa, the last major region to participate in the global dynamism, has seen its growth per capita rising strongly. One sign of the displacement of the center of gravity of the world economy is that when the great recession of 2008–2009 erupted, the financial fragility of the old industrialized nations was fully exposed, with their accumulated debt of hundreds of billions of dollars owed to emerging countries. Would this new recession, forty years after the Club of Rome report, mark the end of an exceptional period of rapid global growth?

The Limits to Growth: The Exhaustion of Resources or of the Capacity for Innovation?

The idea that the world's economies are about to end a long period of growth because they are coming up against the wall of raw materials scarcity has regained ground. Yet historians of growth reveal how societies have managed to elude successive threats by means of technological advances leading to more efficient use of these scarce resources and by increasing their availability through investment in exploration or finding substitutes.

The last forty years do not seem to be an exception. Take the example of oil, without doubt the most iconic commodity. In 1973, an embargo lasting a few weeks gave rise to widespread fears of physical scarcity. But has there been any oil shortage over the last forty years? In 1973, the fledgling International Energy Agency (IEA) estimated that there remained forty years of oil reserves at the extraction rate then prevailing. In 2013, this estimated figure had not changed and, if shale gas and other now exploitable unconventional resources are taken into account, there remain considerably more years of hydrocarbon extraction ahead than there seemed to be in 1973. What is more, since 1973 global GDP has almost tripled, the share of fossil fuels in primary energy production has remained at 80 percent on average throughout the world, and the reserves are still not exhausted.

True enough, the proponents of the peak oil theory will reply, but we cannot extrapolate from the past.[8] Since stocks are finite, there is a physical limit that will one day be reached. We must therefore prepare for this by committing without delay to an energy transition. Conventional economists of energy will add that the physical scarcity of a resource results in price increases that have the threefold effect of encouraging economy of use, facilitating substitutions, and boosting investment in research and exploration. Everyone can see, moreover, how the increase in oil prices has led to a fever of exploitation worldwide, spearheaded now not only by the traditional players of the oil world—multinational corporations and Middle Eastern states—but also by new entrants such as China and Brazil, among others. It is hard to believe that this unprecedented wave of investment will not again lead to further unexpected discoveries in the future.

Though the return to growth is elusive, it is altogether unwarranted to attribute the main cause of this stagnation to the scarcity of raw materials and, by extension, to their projected rise in price. Furthermore, everyone can see the dynamism of the emerging Asian economies—China, Korea, India—which obviously cannot be explained by their domestic availability of raw materials, because in recent decades they have become major importers. Might it not be more appropriate then to consider another resource in accounting for the continuing slowdown of growth in the old industrialized countries—namely the capacity for adaptation and innovation?

Such is the thesis of the economist Robert Gordon, whose analysis forecasts such a decline of growth in the United States.[9] According to Gordon, the potential for U.S. expansion has steadily diminished in recent decades, and the economy will tend toward a near-stationary state with no more than 0.2 percent GDP growth per capita over the coming decades. More fundamentally, Gordon reminds us that "there was virtually no economic growth before 1750, suggesting that the rapid progress made over the past 250 years could well be a unique episode in human history."[10]

Gordon links the closing of this historical parenthesis of the period of growth to the absence of new groundbreaking innovations. In his view, the "anomaly" of twentieth-century growth originated in the changes induced by three major innovations made between 1880 and 1900: the control of electricity (in its production, its distribution, and its uses), the development of the internal combustion engine, and the introduction of running water and sanitation. In other words, three innovation clusters directly related to the new use of existing natural resources.

Gordon argues that these engines derived from innovations made in the late nineteenth century drove the growth of the developed countries until about 1970. Since then, they have tended to lose momentum because the goods and services stemming from these innovations are widely distributed, without any further innovations taking their place. Gordon insightfully shows in his analysis of the digital revolution why the new information technologies do not produce breakthroughs comparable to those engendered by electricity, the internal combustion engine, and general access to drinking water in cities. He also sees the energy and climate constraint as one of the parameters that will permanently be a burden on the U.S. economy, in the same way as rising inequality and debt. Economic dynamism weighed down by climate policies? Bad news for advocates of green growth!

The main theme of the present book is instead based on the idea that the correct approach to natural capital can produce a lasting economic recovery. Where can the breakthrough innovations leading to reinvigorated growth come from? As Gordon shows, they will not emerge from constantly increasing the power of computers or expanding the functionality of smartphones, nor by inventing spacecraft or ultrafast futuristic transportation systems such as those dreamed up by Jules Verne. Taking the environment into account can, however, become the main crucible that will produce these innovation clusters. But the transition to such new modes of growth constitutes a largely unknown path, to which most economic and political actors are reluctant to commit themselves, preferring instead to cling to established frameworks.

To shed light on this innovative potential, it is first necessary to look at how we exploit natural resources—capital that was not originally created by human investment. Viewed naively, this capital is simply a gift of nature in which there is no need to invest because it is constituted independently of past savings and investment efforts. Yet it is clear that in the absence of any investment, it is at risk of being squandered, especially in a world in which the population has exploded, as is apparent when one looks at recent demographic trends compared to secular population growth rates.

2

The Spaceship Problem

An Optimal Population Size?

IF THE LIMITS IMPOSED by natural capital seem to be closing in on the human species, is it not primarily a problem of numbers? *Homo sapiens* is the only species to have expanded its habitat to the detriment of its competitors, many of which have been either systematically domesticated or confined to ever more restricted areas and face increasing risk of extinction. Is there no limit to this occupation of the planet by a single dominant species? In a 1966 article, the essayist Kenneth Boulding introduced this issue using the image of "spaceship Earth" loaded with passengers.[1] This novel concept gave rise to abundant literature by economists seeking to define the "optimal" size of the population. What lessons can be drawn from the literature discussing the question of what constitutes the desirable number of occupants aboard this shared spaceship?

Less and Less Space per Person

From a historical perspective, the figures are staggering. From the year zero to the middle of the last millennium, the world population increased from some 230 million to around 440 million. In 1500, there were still 30 hectares for each person on the planet. Over the last three hundred years the world's population more than doubled to nearly a billion people in 1800, on the threshold of the Industrial Revolution. The average living space per person fell to less than 15 hectares. The nineteenth century saw a demographic surge with the fall in mortality in European countries and their colonies and in Japan. In the twentieth century, this population growth spread around the globe, despite the heavy toll from the two world wars. The rate of acceleration of the global population increased until the 1970s. The six billion threshold was crossed before the end of the century. In 2012, the figure stood at just over seven billion. Living space per person has fallen to less than 2 hectares, against more than 8 hectares a century earlier. According to the projections of UN demographers, it will be down to less than 1.5 hectares by 2050.

Are there not too many people on the Earth in relation to planetary limits, the consumption of resources, and the space available? In general, to try to understand how humans can sustainably coexist, the question of population cannot be addressed independently of people's way of life. The question has long been framed in these terms by both economists and philosophers.

The mercantilism of the sixteenth and seventeenth centuries advocated the largest possible population through various policies. But it should not therefore be assumed that the mercantilists totally disregarded the potential problems of an increased population. In fact, they saw such problems as an argument in favor of migration and the colonization of new territories.

The first reaction to the mercantilist outlook came from the philosopher Giovanni Botero, who explicitly raised the question of the optimal size of the population in his essay *On the Causes of the Greatness of Cities*, published in 1588.[2] Botero highlights the fact that human beings tend to multiply as much as nature allows. However, natural resources are limited and people cannot avoid making certain adjustments: either they change their behavior or there will be automatic adjustment through famine, disease, and war.

In economic thought, Richard Cantillon in 1755 offered another preliminary study of the problem of spaceship Earth, revealing a tradeoff between the "quantity of life" and the "quality of life."[3] This tradeoff leads to a limitation in the size of the viable population. In Cantillon's view, human subsistence requires habitable space (geographic space and economic space), the size of which depends on people's ways of life, so that a tradeoff must be made between quantity and quality. Although Cantillon did not resolve this dilemma, he showed that, contrary to mercantilist beliefs, more people is not always better than fewer people.

Another major contribution to this question was made in 1798 by Thomas Malthus, who argued that the size of the population necessarily has limits.[4] The idea is that in the absence of resource constraint the population naturally grows as a geometric progression, while production of the means of subsistence can at best only follow an arithmetic progression, due to the finiteness of the Earth. The population must therefore, at a given moment, be compatible with the existing space. Malthus then envisages two ways of making the size of the population consistent with that of the Earth: either by "positive" checks (death) or by "preventive" checks (fewer children). Malthus's recommendations regarding population policy are orthogonal to those of the mercantilists.

Historical perspective shows the weakness of any simplistic recommendations to contain population growth because of the limited size of the Earth. When Malthus's celebrated *Essay on the Principle of Population* was published, the United Kingdom had twenty million inhabitants and the planet a billion. Malthus thought that population growth would soon come up against the scarcity of farmland. The United Kingdom now has sixty-one million inhabitants and the planet seven billion, who on average are better fed than in Malthus's time. In the intervening period, humanity has made technical progress unimaginable at the beginning of the nineteenth century. Have we become any more discerning and able to anticipate the technological breakthroughs of the twenty-first century?

In Search of the Optimal Number of Persons

While the positive approaches developed by Botero, Cantillon, and Malthus emphasize the constraints on the size of the population as a result of the finiteness of the Earth, they leave open the normative question of the optimal population. This issue was extensively studied through the utilitarian approach. Jeremy Bentham, the founder of classical utilitarianism, suggested in 1789 that the optimal number of people is that which produces "the greatest happiness of the greatest number."[5] To calibrate this population, he uses the criterion of total welfare, obtained by aggregating the utilities, or levels of well-being, of each individual comprising it. A century later, Henry Sidgwick proposed introducing a second criterion, that of average utility, on the basis that the goal is not to maximize total well-being but average welfare.[6]

These two approaches differ of course with regard to the "optimal" number of people. As Derek Parfit showed in a 1984 book, they lead to unsatisfactory ethical conceptions and raise what is known in the literature as the mere addition paradox.[7] The paradox arises in that the simple addition of individual well-being that is slightly lower than the initial average is viewed as undesirable, even if the welfare of the initial population is not affected.

In other words, if we take the inhabitants of a totally isolated island and calculate the average level of welfare on that island, then the discovery of a population on another equally isolated island is not desirable if the welfare there is slightly lower than that on the first island. We can then see that even in the absence of interaction between the two populations, the principle of simple addition is not a satisfactory criterion. The approach based on averages therefore violates the principle of simple addition, according to which, all other things being equal,[8] adding people with positive well-being is desirable in order to arrive at the optimum population.

In his *Theory of Justice*, published in 1971, the American philosopher John Rawls proposes going beyond this utilitarian conception. He adopts an approach involving welfare but offers a classification of utilities according to the position of different individuals in society. For those who are most vulnerable, he calls into question the conventional rules of additivity. Rawls's approach has fueled the work of economists, who have taken up the idea of a minimum level of welfare that needs to be associated with each individual in society. This approach in fact returns to the problem of the redistribution of resources and wealth among individuals so as to ensure the minimum level of welfare. The main difficulties nevertheless still concern the criteria for classifying utilities and identifying individuals, coupled with a marked lack of consensus. It is evident, therefore, that economists are not able to make specific recommendations in terms of redistribution and ethics with regard to the optimal population target. Thus it seems particularly risky to draw solely on economic doctrine to justify this or that demographic and redistributive policy orientation.

As John Broome pointed out in his 2004 book *Weighing Lives*, one criterion might be that of life expectancy, but the question of the optimal number of people can clearly not be dissociated from that of longevity.[9] Someone who has already been born and someone as yet unborn are, from a moral point of view, quite distinct.

Yet they both exert an influence on the number of people at a given point in time; hence the two ends of the demographic chain are relevant to the study of the spaceship Earth problem.

Given that people want to live long lives with plenty of living space yet have to share a finite space, how should we think about the optimal population in terms of fertility and longevity? To answer the question, we must take into account the effects of congestion, which are likely to reduce welfare as a result of pressures on natural resources. The problem then refers to the choices made by a population aboard the spaceship, by which we understand not only the choices made by the first group of people (the "pioneers") taken on board but also their choices concerning the conditions for their survival, such as individual longevity and fertility, given the finiteness of the spaceship.

Congestion is not unrelated to longevity because overcrowding can prevent access to vital resources (through their depletion or the accumulation of pollution). Thus depending on the greater or lesser initial size of the pioneer group, each individual will have, given the volume of the spaceship, a shorter or longer life. A trade-off thus arises between the number of people and longevity, a tradeoff that demographers agree is the basis for one of the most important concerns of their discipline: the "demographic transition" that results from adapting to lower fertility rates in response to increased life expectancy.[10]

Adjustment of Fertility to Life Expectancy

The demographic transition is most advanced in the industrialized nations. The falling birth rate in these countries has led to a contraction in the base of the age pyramid, foreshadowing a decline in population size in the absence of net inward migration. Given this trend, policies designed to raise the birth rate have had only a limited impact.

The demographic transition has also begun in emerging countries and, with differing time lags, in the least-developed countries. Here, too, the trigger for the decline in fertility has been the improved economic and health conditions associated with better education, especially among women. When demographic policies have been applied on a large scale with a high level of constraint, as in China with the one-child rule, they may have accelerated the spontaneous decline in fertility that would have occurred anyway. Their effectiveness from a long-term growth perspective, however, is questionable: the overly rapid narrowing of the base of the age pyramid will result in China having to face a sudden aging of its society in coming decades, combined with a population decline estimated by UN demographers at about fifty million for the period from 2040 to 2050 alone.

This observed link between increased life expectancy and reduced birth rates undermines the basic premise of Malthus and the whole current of thought he initiated. If the birth rate reacts, with a lag of a generation or two, to the increase in life expectancy, population growth deviates from the geometric progression described by Malthus: the slower pace of population growth slows the increase in the number of occupants in the spaceship. The postulated geometric shape of its increase may even become a logistic function (an inverted U-curve) if the rate of population growth tends to zero in the long term.

It is this curve that UN demographers foresee in their projection of the world population. Based on historical observations, the rate of increase of the inhabitants of spaceship Earth peaked in the 1970s at more than 2 percent per year, or a doubling of the population every thirty years. As a result of the observed decline in fertility, this rate fell to 1.2 percent during the first decade of the twenty-first century. In 2050, it is expected to be 0.4 percent, with a ship's complement of some nine billion people. At this rate it would take 170 years for the population to double. But demographers foresee a continuing decline in fertility leading to near-stagnation of world population at a little over ten billion at the end of this century.

While it is indisputable that a population of ten billion people puts more pressure on natural resources, at comparable living standards, than a population of 230 million (estimated for the year zero), Malthusian recommendations in terms of population limitation or decrease do not seem relevant. The search for an optimal population size is fraught with conceptual difficulties that neither the utilitarianism of economists nor the quest for an equity rule by moral philosophers will resolve operationally. The effectiveness of population policies to encourage or hinder the birth rate also seems very low from the standpoint of the evolution of life expectancy.

To avoid famines, Malthus, an Anglican clergyman by profession, recommended that the poor refrain from having children. He seemed to be unaware that the problem of British society at the time was precisely the number of low-waged people generated by the early stages of industrialization. The question of the number of people that can be accommodated aboard the spaceship in the twenty-first century is no different. It depends primarily on the capacity of societies to further disseminate wealth and education among their populations. In other words, it raises the question of the distribution of the fruits of growth, which is one of the least discussed but most strategic aspects of the ecological transition.

3

Degrowth

Good Questions, Bad Answers

ANOTHER WAY OF EXTENDING the voyage of the spacecraft is to act not on the number of occupants, as Malthus recommended, but on the average wealth of each occupant. This is the option explored by advocates of degrowth. On behalf of protecting the environment, should we be moving toward green degrowth?

This option was also considered early on by the classical economists. In 1848, fifty years after the publication of Malthus's *An Essay on the Principle of Population*, John Stuart Mill published *Principles of Political Economy*. In this work he moves away from Malthusian pessimism, in particular the assumption that the means of subsistence has little growth potential. Doing so did not prevent Mill from showing great caution with regard to growth and extolling the virtues of a "stationary state." Unlike Malthus, Mill did not resign himself to this stationary state because of the scarcity of cultivable land. Instead he was in favor of it in the interest of seeking the welfare of the community. Indeed, in this stationary state, "there would be as much scope as ever for all kinds of mental culture, and social and moral progress; and as much room for improving the art of living, and much more likelihood of it being improved, when minds cease to be engrossed by the art of getting on."[1] Mill's thinking is premonitory. Might he be viewed as the first "growth objector"?

Degrowth as an Antidote to Technological Optimism?

Degrowth is at once a current of thought and a protest movement against the capitalist system. Its theoretical foundations were laid after World War II by the Romanian-born economist Nicholas Georgescu-Roegen. In his opinion, the physical law of entropy, stating that any closed system evolves only by becoming slightly more disorganized and making part of the energy it contains unusable, also applies to our economic system. In the long run entropy implies that the economic system is doomed due to the depletion of energy and material flows. He therefore advocates economizing these flows through degrowth in order to gain time with regard to the inevitable outcome.

Thus formulated, the doctrine of degrowth recalls Keynes's well known dictum "in the long run, we are all dead"—a useful reminder for the conduct of our personal lives but a little offbeat for the analysis of economic systems.

Closer to our concerns, the proponents of degrowth, represented in France by authors such as Serge Latouche, Jean Gadrey, and Paul Ariès, point out that much of the gain in environmental efficiency is "eaten" by growth. In 2010, the average-sized car in France consumed half as much fuel as in 1970—a remarkable enhancement of performance. But the environmental benefit was more than offset by the increase by a factor of 2.4 in the number of cars over this period and by the 20 percent increase in mileage per vehicle. There are many other examples of this type of "rebound effect."[2] They all point in the same direction: the amount of energy, greenhouse gas emissions, and environmental damage involved in producing each unit of GDP is decreasing (at least in the advanced economies), but the effects of this environmental decoupling disappear due to the increase in the wealth produced. To escape this impasse, orthodox economists argue that the pace of technical progress needs to be speeded up. Proponents of degrowth retort that such blind faith in progress imprisons orthodox economists in an irresponsible form of "technological optimism" and that the right route should be to reduce the amount of wealth produced.

Proponents of degrowth question the benefits of technical progress. Georgescu-Roegen did not suppose for a moment that greater knowledge leads to innovations that can significantly delay the laws of entropy. Following him, growth objectors question technological optimism and its long list of revolutionary promises: nuclear energy, long promoted as abundant, safe, and able to "permanently" solve the energy problem; miracle seeds derived from genetic engineering, an unexpected remedy for world hunger; thin layers of future-generation photovoltaic cells to convert sunlight into electricity that can be attached to houses or even wrapped around clothes like a coat; miracle plants that will produce clean biofuels that do not compete with food production; incredible geo-engineering systems that will directly prevent global warming—for example, by protecting the Earth through an array of giant mirrors reflecting sunlight or by seeding the oceans with iron sulfate to increase their carbon-absorbing capacity.

This technological optimism often accompanies the failure of policy makers faced with a problem they do not know how to or do not dare tackle. Is it any surprise that there is a renewed interest in geo-engineering when the prospect of an agreement on climate change is receding? As for eliminating hunger in the world through genetically modified seeds, that is sheer madness: malnutrition is the result not of insufficient food production capacity but of poverty. Whether harvests are good or bad, if the prices of staple foods soar as they did in 2008, the number of malnourished people increases dramatically (one hundred million more in 2008, according to the FAO and the U.S. Department of Agriculture).

Another advantage of technological optimism is that any calling into question of the current model of economic development can be evaded. Its adoption allows us to imagine that the world of tomorrow will be an extension of today's, but with the introduction of technological breakthroughs miraculously solving environmental issues. The IEA puts forward production and energy consumption scenarios for 2050 with much less CO_2 emitted into the atmosphere thanks to technological breakthroughs. But what social changes will be entailed? Which people will gain and which will lose out? What social and political forces will be needed to create the conditions for such advances? Advocates of degrowth have been quite right to criticize this approach, in which technology is a deus ex machina and miraculously solves all problems.

For all that, is green degrowth the right antidote? Censure of technological optimism by its critics is likely to lead to a decrease in research and development efforts. Yet the transition to a green economy requires a proliferation of innovations that do not fall from the sky but instead require increased R&D investment. This applies not only to industry and services but also to agricultural activities.

From a degrowth perspective, agriculture would become more environmentally responsible by turning to more extensive exploitation, particularly the reduction or even elimination of the use of chemicals. It would almost amount to a step backward to techniques used in the past. But given the demographic situation, orienting agriculture toward such extensive methods would be a surefire way to increase the pressure on non-cultivated areas, starting with the most accessible woodlands and wetlands, which are valuable reservoirs of carbon and biodiversity. The future of organic or environmentally responsible farming systems lies, therefore, in intensification based on new technical procedures. The agronomist Michel Griffon has clearly shown how the development of such "ecologically intensive agriculture" calls for considerable investment in research and development.[3] For farmers and stockbreeders, adopting green techniques based on the complementarities among the different elements of natural capital is infinitely more complex than implementing the fertilizer and pesticide application standards provided by the local cooperative.

Reducing Non-Essentials and Waste: The Case of Junk Food

Another pertinent analysis by the proponents of degrowth concerns the alienating forms of consumerism that trap people in an "iron cage"—the term used by Tim Jackson in one of the most persuasive critiques of contemporary growth patterns.[4] Jackson brilliantly exposes capitalism's capacity to amplify consumer needs artificially, which expands the mass consumption market without increasing people's welfare, and even subjects them to unhealthy frustration through not being able to climb to the heights of consumerism. Dissatisfaction then becomes one of the drivers of consumption and the market becomes a gigantic machine to create this dissatisfaction.

One does not have to look far to find thousands of examples of such superfluous purchases of goods of dubious utility and rapid obsolescence. But one area in which the costs of this type of waste are particularly shocking is food.

In orthodox economic thinking, food items are viewed as necessities, that is, goods for which the demand responds little to price changes.[5] In addition, food consumption cannot in principle increase indefinitely because of the "stomach wall": once satisfied, the consumer should spend his or her budget for other purchases. Consequently, one expects to see the share of food in household expenditure decline, though in practice this reduction is not seen in consumer surveys (especially if meals outside the home are taken into account). Indeed there are many ways to enrich the diet, particularly by incorporating animal products, which are much more expensive to produce. Moreover, food consumption can always involve more processing, transport, packaging, and storage. The intake capacity of the stomach in no way limits the thousands of kilometers food travels or the time it spends frozen. It is therefore necessary to enrich conventional economic thinking in order to understand the dynamics of these markets and to transmit the right incentives.

Historically, the diversification of diets originally consisting of a limited number of inexpensive starchy or grain-based foods greatly contributed to the increase in life expectancy and welfare. But this transition from scarcity to abundance has generated new problems. The pursuit of richer food, with overconsumption of meat, fats, and sugar in high-income societies, has become a major source of ill health and mortality. The prevailing dietary habits are producing obesity, diabetes, cardiovascular diseases, and all sorts of other health problems. At the other end of the chain, they are putting increasing pressure on resources because they require much more area, water, and energy to maintain supplies. At the same time, requirements for food aid are increasing year by year, as noted, for example, by the charity Community Ministries, because the widening of the range of incomes makes this model unattainable for those people at the bottom of the ladder.

Reducing non-essentials, ostentation, and waste is clearly a desirable objective. In the case of food, there is even an additional public health goal. But who will determine the boundary between what is useful and what is superfluous? And what possible methods could be adopted to make the latter decrease and the availability of the former increase? The time when a centralized planner was trusted to dictate what the consumer should or should not buy is over. Some people, like Jean Gadrey, call for participatory planning, but apart from a quasi-ideological calling into question of conventional notions of wealth and growth, its linkage with representative democracy has yet to be specified and its decision-making criteria are not always explicit.

In fact, it is not clear how this type of approach would concretely help society guide eating habits toward choices that reduce waste in terms of both demand and supply. The method we propose here is to analyze the current functioning of markets to improve them through a series of actions designed not to block growth but to reorient it by means of new incentives. Four rules are adopted: 1) the markets should be open, transparent, and convey unbiased information; 2) they should price all environmental and health-impacting pollution; 3) they are inappropriate to replace the public authorities regarding infrastructure choices that affect ways of life; and 4) they should not be allowed to dictate the rules of income distribution and access to essential goods. Let us briefly consider see how these four rules might apply to food.

1) The quality of information, particularly nutritional information, transmitted to consumers is a major consideration. It is related to the structure of markets in which private operators dominating certain segments can convey messages altering behavior in a way that furthers their interests but is harmful to public health. For example, U.S. studies on obesity reveal the influence that fast food operators and soft drink manufacturers have had in the United States.[6] Furthermore, the opening of markets to new short distribution chains is a prerequisite for moving toward the greater diversification of diet recommended by nutritionists.

2) Markets should correctly price the environmental and health costs of food. As the markets are currently organized, the environmental costs of a side of beef produced on land cleared of forest in the state of Para in the Amazon basin, frozen at a facility in the Port of Belém, shipped to Rotterdam, and served in a hamburger in Maastricht are not included in the price of the product. The pricing of food items also fails to reflect health priorities. Everyone can see that high-fat products have become cheaper than low-fat products on supermarket shelves, thereby sending a signal diametrically opposed to healthy eating. The nutritionist Marion

Nestle has calculated that it would cost each American $400 a year just to comply with federal nutritional recommendations.[7]

3) The gap between the quantity of food we ingest and our nutritional needs is also related to lifestyles that limit our physical activity. Here it is also a matter of infrastructure influencing behavior: in the United States, for example, obesity is more prevalent in fragmented cities where people spend a good part of the week sitting in their vehicle. The other major factor in weight gain is the proliferation of TV and computer screens, coupled with the long hours people sit in front of them, very often with a snack within easy reach! Transport and communication infrastructure plays a major role in the rise of nutritional problems.

4) The overabundance of food items has not succeeded in eradicating want. Due to growing inequality, the income pyramid is extending at both top and bottom. The lives and conspicuous consumption of the super-rich are paraded in celebrity magazines, while the destitute line up at soup kitchens. This rising food inequality is also found, on an even larger scale, in developing countries. A country such as India is faced simultaneously with undernutrition affecting hundreds of millions of rural people and growing health problems caused by the food overconsumption of the new urban middle classes. Specific instruments are needed to address this issue of equity; it should not be left up to the markets.

Degrowth in the North for Improved Growth in the South

Criticism of the damaging effects of growth is easier in the advanced economies than in the least-developed world, where most of the population does not have access to essential goods, or even in emerging countries, where efforts are being made to benefit the greatest number. It is for this reason that an author such as Jackson advocates halting growth only for high-income countries. Even better, he believes, would be to envisage a set of communicating vessels in which degrowth in the North becomes a lever for development in the South: "A key motivation for rethinking prosperity in the advanced economies is to make room for much-needed growth in poorer nations."[8]

If halting growth in the high-income countries automatically boosted growth in the poorer countries, there would be little hesitation: it would be a matter of rapidly introducing a growth reduction policy in the advanced economies, while explaining this to the electorate so as to obtain its agreement. First problem: How would the cost of any such degrowth be shared? Without a powerful corrective policy, it is a safe bet that it would not be the elite but the poor who would suffer the most—that is, those on behalf of whom, at the international level, degrowth was justified.

But suppose the first problem can be resolved. The second problem is then: Up to what point should growth credit be given to low-income countries? And who decides how much? The precedent of climate negotiation is worth thinking about. According to the principle of "common but differentiated responsibility" under the United Nations Framework Convention on Climate Change signed in 1992, "rich" and "poor" countries were distinguished on the basis of whether or not they belonged to the OECD in 1990. As a result, countries such as Singapore and South Korea were classified in 2014 as poor, while Greece and Bulgaria were included in the group of rich countries. Every climate negotiator knows how much emerging countries are opposed to any questioning of this historic demarcation by representatives of the old rich countries!

The truth is that the growth process is based on a complex set of interdependencies that disseminate innovations, open up markets, and support investment. If one of its main engines is damaged by a growth interruption policy in the North, the effect will be a severe setback in the South and not the hoped-for outcome from the interplay of communicating parties. The real issue raised by Jackson is the rising inequality that has accompanied policy choices in recent decades. These inequalities are now viewed as an obstacle to further growth, not only by the American economist Robert Gordon but also by a journal as little suspect of indulging left-wing thinking as *The Economist*.[9] As we will see in later chapters, the transition to a green economy can increase or reduce these inequalities, depending on the choices made.

While the degrowth approach raises good questions, it does not provide satisfactory answers. Worse, it risks trapping society in a false dilemma: either the environment or growth. In the context of the economic crisis and rising unemployment experienced since 2008, there is little doubt as to their respective priority. Instead what should be aimed for is a redeployment of growth that incorporates the objectives of protecting natural capital and stimulating research, investment, and innovation. On behalf of the environment, we must not halt growth but change it.

4

Introducing the Environment into the Calculation of Wealth

SEPTEMBER 14, 2009. A busy day for the Sorbonne. In the packed grand auditorium, a succession of Nobel Prize winners in economics is addressing the audience. In the first row, a particularly attentive listener may be seen: none other than Nicolas Sarkozy, president of the Republic. In an impressively orchestrated economics-media operation, the Stiglitz-Sen-Fitoussi Commission is presenting its findings on the "Measurement of Economic Performance and Social Progress,"[1] a weighty document of more than 300 pages (with appendices) that the Commission's rapporteurs have taken a full year to produce. Over the coming days, the media covering the event, seeming suddenly to have developed a passionate interest in the nation's wealth accounting procedures, will discover that the celebrated GDP is not an indicator of human welfare. By changing the instrument for measuring growth and incorporating sustainable development objectives—the boldest commentators have no hesitation in asserting—economic policy will spontaneously be enriched.

The addiction to GDP growth has been particularly well analyzed by the sociologist Dominique Méda,[2] who wants to find ways of freeing society from this collective mystique. For its part, the public cares little about how GDP is calculated. On the other hand, it is well known that the growth of this indicator is directly correlated to two of people's main concerns: employment and purchasing power. It is for this reason that the addiction to GDP is unlikely to diminish, regardless of the number of Nobel Prize winners gathered at the Sorbonne to discuss its limitations.

In the intervening years, it has become clear that the Stiglitz-Sen-Fitoussi report has not had the desired impact. It had the merit of reminding us that the GDP growth in no way measures any increase in social welfare. In terms of the environment, it testifies to the progress made in the fight against local pollution and to the severe deterioration of the major regulatory functions of the climate and biodiversity. But with regard to incorporating these considerations into economic life, the real revolution would be to price the costs of environmental damage and include them in our microeconomic accounts—which we use to make our decisions.

Wealth and Welfare

The Stiglitz-Sen-Fitoussi report was a comprehensive synthesis of existing studies. But did it introduce any major innovations? Concerning the relationship between the calculation of GDP and the calculation of welfare, it draws on the contributions of the seminal paper by the economists William Nordhaus and James Tobin in the 1970s.[3] In this they argue that welfare largely depends on values not included in GDP, such as the production of domestic services or the strength of social ties within a community. Symmetrically, because GDP is based on flow accounting, it includes in the calculation of wealth the value of the expenditure required to repair damage after a natural disaster or a war. Tobin and Nordhaus subsequently proposed calculating two indicators of welfare, the first on the basis of household consumption, the second introducing a measure of depreciation of natural capital.

Following their work, many complementary or alternative indicators to GDP have been developed for measuring welfare, incorporating both social and environmental dimensions. The most well known is probably the Human Development Index (HDI), calculated by the United Nations Development Programme (UNDP). This indicator has been published annually since 1990 for all the world's countries, to complement the ranking of countries based on GDP. Not surprisingly, we find that for oil states, such as Qatar and Angola, or countries where there exists very great inequality, such as South Africa, performance in terms of the quality of life and welfare lags well behind their GDP level. Conversely, the Nordic countries, and to a lesser extent France, perform better than GDP thanks to their social safety nets. The statistics thus corroborate intuition. Yet the development policies implemented around the world have not been changed following the dissemination of these new indicators.

Another alternative measure came with the development of a Genuine Progress Indicator (GPI), which incorporates the estimated cost of present and future environmental damage resulting from economic activity and takes into account debt and social inequalities. Observation of this indicator in the United States reveals a disconnect between the evolution of GDP per capita and that of GPI per capita: from the early 1980s, the former has continued growing while the latter has fallen slightly, particularly because of rising environmental costs (mainly the consumption of exhaustible resources and greenhouse gas emissions). These costs represent a levy of 60 percent when charged to household consumption as measured in the national accounts. Other factors accounting for this dramatic decoupling of these two indicators in the United States are rising debt and growing inequality, which have impacted the growth of welfare without affecting GDP growth.

Regarding the environment, it has become the norm in developed countries to complement the macroeconomic accounts with satellite accounts that allow the impact of GDP growth on the state of the environment to be assessed. The Stiglitz-Sen-Fitoussi report recommends assembling a set of physical data in the form of a "dashboard" whose nomenclature has yet to be specified. It stresses the difficulty of constructing composite indicators and the even greater difficulty of monetarily introducing them into a national accounts system. Let us now take a closer look at this environmental component.

As rightly pointed out by Stiglitz, Sen, and Fitoussi, the problem for those concerned about ecology is not so much the shortage of information as its surplus. With a view to raising public awareness, specialized international organizations regularly produce comprehensive balance sheets on the "state of nature," usually backed up by projections. An excellent summary is provided in the book by Edward Barbier and Anil Markandya, *A New Blueprint for a Green Economy*,[4] which has now been updated twenty years after the first edition, coordinated by the late David Pearce. These authors carefully distinguish changes in local pollution and global environmental damage.

Local Pollution: Uneven Progress, Positively Related to Wealth

The record of the past three decades is mixed regarding local pollution. The air quality in cities, as measured by the concentration of microparticles, has improved by more than a third over the last quarter century. Progress has been much faster in the cities of high-income countries, widening the gap with the countries of the South (where there is maximum pollution in the cities of those countries that have just lifted off from the lowest poverty levels). Another important indicator of air quality, the sulfur dioxide (SO_2) content, has sharply declined in high-income countries where the fight against acid rain has been a priority, whereas it has increased in many developing countries, such as China and India.

The same contrasting picture applies to pressures on water. The water quality of rivers has generally improved in developed countries (like the Seine in Paris), in contrast very often to developing countries, where the deteriorating water quality of rivers causes major problems (as in India, with the pollution of the Ganges). Although rapid progress has been made over the last ten years in terms of the poorest populations' access to fresh water, this is not the case with regard to the treatment and disposal of waste water. In quantitative terms, the number of people living in conditions of "water stress" has increased sharply since 1970, and thirteen countries, all of them in the developing country category with the exception of Israel, take more water every year from their hydraulic resources than they put back.

This uneven progress, depending on the wealth of the countries concerned, brings to mind the bell shape of the so-called Kuznets curve introduced in 1995 by Grossman and Krueger.[5] These two authors attempt to relate environmental damage to economic development by finding a similar type of relationship to the one Kuznets established between development and economic inequalities: in the early stages, the growth of a country's wealth is accompanied by increased pressure on the environment, with negative consequences on natural resources and public health. A peak is eventually reached, at which point pollution starts to decline. Grossman and Krueger do not claim any deterministic relationship, but simply link this bell-shaped curve to shifts in the population's values. Beyond a certain level of wealth, the utility attached to food and manufactured goods decreases, while the perceived utility of environmental goods increases. Households are then increasingly willing to pay to protect their environment through public policy. A similar pattern also seems to apply to deforestation mechanisms: in recent decades, boreal and temperate forests located in high-income countries, which were historically the first to clear primary forests, have expanded; at the same time, the tropical forests of Latin America, Africa, and Asia have suffered accelerating deforestation.

Observation of traditional local pollution and of deforestation tends to corroborate the Kuznets environmental curve and support the idea that at a certain level of wealth, societal changes lead to environmental issues being better managed. But this sum of partial views is misleading. Firstly, it underestimates the possible relocation of certain polluting sources and industries in response to rising environmental concern in the North. Secondly, and most importantly, it fails to take into account the damage caused by our current patterns of growth to the two great regulatory systems that ensure ecological diversity and the stability of the climate system.

Disruption of Regulatory Systems: Biodiversity and Climate Stability

The diversity of living species is a major feature of ecosystems that provide a variety of "free" services to human economic activity by promoting the regeneration of resources. Several waves of extinctions have been identified in past geological eras, the most well known (but not the most important in terms of biodiversity) occurring on the threshold of the Tertiary period, with the extinction of the dinosaurs sixty-five million years ago. These mass extinctions have generally been linked to climate changes and/or exceptional phenomena involving volcanic eruptions or meteorite impacts. A new wave of extinctions has started within a historical period that is very short in comparison to geological eras. Unlike those of the distant past, this one has an entirely new cause: the domination of one living species, mankind, which in the course of its growth has eliminated and domesticated many competing species.

Measurement of the damage to the diversity of ecosystems leaves no doubt about the scale of the phenomenon: our historical epoch is producing a much more rapid extinction of living species than those of past geological eras.[6] The most publicized indicator, the Living Planet Index (LPI) calculated by the WWF, reveals a decline of almost a third of the diversity of 800 populations covering 2,500 vertebrate species since 1970. Over the twentieth century, the global LPI declined by 21 percent, of which 12 percent was accounted for by the period from 1990 to 2000 alone. No area of the world has been spared, though the LPI for temperate forests, in contrast to the general trend, has been increasing in recent years.

Faced with this phenomenon, the consequences of which are multifaceted and the associated risks high, the first corrective actions taken involved developing protected areas, of the kind classified as Natura 2000 in Europe. The idea is to create "reserves" of biodiversity, kept apart in varying degrees from areas of human economic expansion and its damaging effects. It is clear that these forms of defense are in the long term about as effective as building dikes to protect coastal cities against rising sea levels without acting on its causes. One of the key imperatives of the ecological transition is to find the right incentives to integrate the value of the diversity of life into the functioning of the economy. As we shall see in more detail, this is complicated by the many ways in which biodiversity is adversely affected and the difficulty of finding a common metric.

Global warming is one of the levers through which human action alters the conditions of reproduction of other species. Warming could in the longer term become a major cause of loss of biodiversity in various ways. First of all, it disrupts existing habitats by fragmenting them and driving many species farther north or to higher altitudes: over the last twenty years, it is estimated that the ranges of European butterflies and French breeding birds have shifted more than 200 kilometers northward.

The increased frequency of extreme weather events linked to global warming has an often overlooked impact on biodiversity: some of the most dangerous events for river ecosystems are violent summer storms, which leach the soil and sweep polluting agents of all kinds into rivers. The risk of wetlands drying out threatens some of the world's richest ecosystems, especially Amazonia and parts of western Africa. Lastly, we should mention the risk of changing distribution of pathogenic agents—for example, the northward spread of dengue and malaria—and the threat to marine ecosystems from the acidification of the oceans.

Climate risk is undoubtedly the environmental concern for which awareness has come to predominate over the past three decades. Unlike for loss of biodiversity, action against climate change has a common standard: the warming power of a ton* of CO_2 released by humans into the atmosphere (and comparable measures apply to other greenhouse gases). The existence of such a standard has facilitated the introduction of emissions pricing instruments, through taxes or tradable allowances, both at international and national (or European) levels. Terms such as "carbon price" or "carbon value" have now entered common parlance. However, it is important to appreciate that this pricing mechanism applies to only one of the major regulatory functions of natural capital. Building a strategy for the transition to a green economy based solely on the price of carbon could lead to serious disappointments, particularly in terms of biodiversity. For example, only taking into account the energy value of a forest, with a view to substituting biomass for fossil fuels, can lead to exploitation methods that are extremely damaging to the diversity of forest ecosystems.

After this quick overview of environmental pressures, it is tempting to look for an overall indicator. Whether it is a matter of the atmosphere, water, biodiversity, reserves of exhaustible resources,

---

*The term "ton" and compounds such as "gigaton" are in metric units throughout (1 ton = 1,000 kg).

or climate change, the basic data comes in the form of physical quantities and therefore cannot be directly aggregated. The concept of an "ecological footprint" developed by William Rees and Mathis Wackernagel seeks to provide an aggregate measure on the basis of a unit of area, the hectare.[7] The idea is to calculate the area needed for the population using a territory to renew its natural capital indefinitely. By aggregation, we obtain the overall footprint that measures the total renewable resources levied on ecosystems for the direct consumption of materials (agricultural, fibers, water, etc.), the recycling of their waste, and the absorption of their pollution. This footprint can then be compared to the physical biocapacity of the planet. The evolution of this indicator has been popularized by the WWF, which regularly publishes it in its "Living Planet Report." The WWF's key message is that the consumption of resources as measured by the indicator has exceeded the biocapacity of the planet since the mid-1970s. According to WWF figures, we are currently consuming about 1.6 planets.

Closer observation of the indicator reveals, however, that most of the increase in the ecological footprint stems from greenhouse gas emissions, which are converted into hectares on the basis of a highly simplified coefficient measuring the atmospheric CO_2 absorption capacity of the soil and forests. We would arrive at very different conclusions by modifying these coefficients or using other aggregation systems. One major difficulty in establishing indicators expressed in physical quantities lies in how to define a common unit and an aggregation system.

Introducing the Value of Natural Capital into Microeconomic Accounting

The most common way to aggregate environmental costs and benefits is to assign them a monetary value using a method that in principle is very simple. Natural capital may be considered part of the stock of capital used in production in the same way as capital goods, available technologies, and human knowledge. Variation in this stock represents the changes occurring between two "states of nature." It is obtained by aggregating inflows (improved regenerative capacity of natural systems, investment in environmental protection) and outflows (overexploitation of resources, impairment of the regulatory capacities of natural systems). In economic terminology, inflows are "investment" and outflows "depreciation."

Since 2004 the World Bank has produced a Genuine Savings Indicator for countries around the world, calculated by incorporating the value of these investment and depreciation flows of natural capital and human capital into national accounts. The method is applied to savings, which in national accounting is the complement of consumption in relation to income. The particular advantage of this indicator is that it clearly exposes the illusion produced in the short term by the capture of resource rents. For example, a country like Saudi Arabia, which posts an apparently high savings rate in its national accounts, is actually in a negative real savings position, reflecting the ephemeral nature of its growth pattern. On the basis of the Genuine Savings Indicator, France had a real savings rate of only about 8.5 percent in 2010 (compared to about 20 percent according to the conventional gross savings rate), and the United States had almost zero.

We can now easily find the elements needed for greening the calculation of GDP in the satellite environmental accounts. However, these elements remain on the periphery of the central frameworks of national accounting, which continue to favor GDP as the indicator of wealth. This choice is justified by the fact that the standard national accounting system is based on the same conventions as microeconomic accounting, which is a standardized and binding framework within which all economic actors evolve. The role of national accounting is therefore simply to aggregate measures of wealth given by multiple microeconomic accounts, which also serve to establish the tax base.

The method is disputed by many environmentalists and alter-globalization organizations, some of which are actively campaigning for the redefinition of the concepts used to calculate GDP on the grounds that it gives a distorted view of wealth. In doing so, they are probably engaged in the wrong fight. The real issue for the construction of new indicators is situated more at the microeconomic level. Rather than talk about the limitations of the GDP indicator, it is much more useful to gradually bring the value of natural capital into microeconomic accounting, used in day-to-day choices in managing businesses and making decisions. This can be done in two ways: by including measures of natural capital in the balance sheet or by assigning a price to environmental values in the market for goods and services. Once the value of the environment is included in their profit and loss accounts, actors' choices change, and the problem of macroeconomic indicators no longer arises. When the damage of pollution or overexploitation of ecosystems is correctly attributed, statisticians integrate it into the national accounts without even realizing it.

Such a move would completely change the way economists analyze the links between the environment and the economy. It would alter the standard representation they have of natural capital.

5

"Natural Capital" Revisited

EVER SINCE THE BEGINNING of the Industrial Revolution, successive voices have claimed that the growth process will be brought to a halt by scarcities of various kinds: arable land, fossil energy, water, fish stocks, metals extracted from "rare earths,"[1] etc. The biggest names of political economy were not averse to making predictions that would subsequently be controverted by the facts. In the late eighteenth century, Malthus announced that population growth would come up against the scarcity of farmland. David Ricardo, in *On the Principles of Political Economy and Taxation* (1817), argued that the limits of the Earth—finite in size and with decreasing fertility—imposed a limitation on the growth of the economy. William Stanley Jevons contended, in *The Coal Question* (1865), that the depletion of coal, combined with population growth and the lack of substitutes, would lead to the same outcome. A little over a century later, the Meadows Report (1972) again sounded the alarm, stressing the physical limits to growth in terms that would not have been disavowed by Meadows's eminent predecessors. This long tradition is based on the widespread view of natural capital as a stock of exhaustible resources. Yet to properly take account of contemporary environmental issues, economists must go beyond this standard view and adopt a more systemic representation of natural capital understood as a set of regulatory systems.

The Standard View of Natural Capital in Economics

In all cases, the reasoning is based on a shared view of natural capital defined as a set of scarce resources, renewable or non-renewable, whose availability may be endangered by the type of development being pursued.

Those opposed to the defenders of the depletion thesis base their arguments on the very same concept of scarce resources. One of the most celebrated critics of these theories is the economist Robert Solow, originator of the neoclassical approach to growth theory. In various papers, Solow argues that assertions such as those made by the Club of Rome underestimate the possible substitutions in the production process that will, if a scarce resource is lacking, lead to investment in another resource by way of replacement.[2] As discussed in more detail in chapter 3, the Solow approach has been taken up by various economists who show how, on the strength of such substitutions, one can envisage infinite growth based on finite natural resources. This controversy around the greater or lesser substitutability of natural capital has given rise to various forms of the concept of sustainable development in the strong or weak sense. The controversy is difficult to resolve as it stands because it draws on a standard view of natural capital defined as a simple aggregation of scarce resources.

Yet this view seems increasingly at odds with the rise of contemporary environmental issues. Consider the example of oil and gas resources, introduced in chapter 4. It is clear that in 1973, as in 2013, the debate among experts on exploitable hydrocarbon resources was nowhere near reaching agreement. In fact the debate turns on the time frame envisaged: the proponents of the doctrine of peak oil believe the depletion of oil resources is imminent, while their opponents focus on new resources, conventional or otherwise, that can extend the lifetimes of oil and gas. In this debate on the amount of potentially exploitable energy from underground deposits, one dimension seems to have been totally overlooked: the impact on the climate system.

Absent from people's concerns in 1973, this dimension emerged as central to the debate from the 1990s onward. It is no longer a question of the risk of the exhaustion of this or that energy source, but of the capacity of the atmosphere to continue regulating energy flows entering and leaving the planet through the greenhouse effect. Within this new perspective, the risk of depletion of fossil resources is less worrying than excess CO_2, which, by building up in the atmosphere, alters the major regulatory function provided by the climate system. Concomitantly, we have gone from a view of natural capital defined as an aggregation of scarce resources—economists' traditional definition since Malthus—to a view in which this capital is a complex whole ensuring the reproduction of resources. With regard to energy, the climate system plays a central role in the long cycle, in which ultimately all the energy we use comes from the sun. Advances in climate science show that disruption of this regulatory function risks affecting, through the impacts of climate change, the resilience of ecosystems by threatening their diversity. The climate system is therefore only one component of the overall regulatory mechanisms ensuring the reproduction of resources.

Natural Regulatory Systems Weakened by the Race for Raw Materials

The example of oil, some will say, is biased. Not necessarily! Shaped by economists, the current view tends to reduce questions that concern the functioning of natural regulatory systems to issues of scarcity and cost. People are alerted to the decline in the number of pandas in Asian forests or of fish stocks in oceans, but without the real issue being clearly identified, namely the degradation of the ecosystems that allow these species to reproduce. Is water becoming scarce and increasingly expensive? The total amount of water available on the planet is the same as it was millions of years ago, but our development tends to increasingly disrupt its natural cycle, sometimes drastically reducing its accessibility and quality. Is there a risk that there will not be sufficient wheat or rice to feed tomorrow's nine billion people? New investors will appear, ready to buy up and clear millions of hectares in response to this ancestral fear of a shortage of agricultural land. In so doing, they destroy forests and wetlands playing a vital role in biodiversity—biodiversity that agriculture very much needs if it is to produce in a sustainable way.

In all these examples, starting from the standard view of natural capital defined as the sum of scarce resources provided by nature, we soon see not only that this capital cannot be reduced to a simple aggregation of scarce resources, but also that it involves a set of regulatory systems. These systems determine in particular the reproduction of resources that used to be seen as "free," that is to say, exploitable in an indiscriminate and limitless manner: water, air, climate, the diversity of life, and so on. To reproduce this natural capital, care must be taken not to exhaust scarce resources if substitutes for them have not been found. But it is also essential to ensure that societal development does not alter the main regulatory functions inherent in this capital, particularly in view of the many efforts made to shift the traditional limits pertaining to the scarcity of raw materials.

Yet if human ingenuity is managing to hold the physical limits of growth at bay, this does not mean that it is contributing to the preservation of natural capital viewed as a global regulatory system. Beyond certain thresholds, increasing investment in the exploitation of raw materials seems instead to be at the expense of these regulatory functions. We have seen this for climate change. It also applies to biodiversity. As Bernard Chevassus-au-Louis, one of the leading French experts on this issue, has noted, "When one alters an ecosystem to increase the marketable products obtained from it—for example, the development of shrimp farming in tropical coastal zones or the exploitation of forests to produce quality timber—, the total value of services provided by their biodiversity decreases, often to a considerable extent."[3]

Over the past four decades, planet-wide economic development has augmented mankind's capacity to push back the limits to growth in terms of the depletion of natural resources. Price increases in world markets periodically alert actors to these risks and give rise to investment strategies to ensure the security of supplies, in which emerging countries now play a pivotal role. As increased climate risk and threats to biodiversity show, this race for raw materials multiplies the methods deployed for exploiting these resources, to the detriment of the vital regulatory functions of natural capital.

In sum, historical analysis offers a way of describing natural capital not as a mere aggregation of finite resources, but as a set of regulatory functions that some types of growth can disrupt and others can strengthen.

Toward a Systemic Approach to Natural Capital: The Bases of a Circular Economy

How is this set of regulatory functions to be defined? Following the work of the Stockholm Resilience Centre,[4] we can identify nine "planetary boundaries" in which scientists have determined there to be a risk of abrupt and irreversible environmental change. The idea is that, beyond a certain threshold always difficult to estimate, one or another of the major regulatory functions of the natural system is no longer assured. Crossing this threshold then weakens the reproduction of all natural capital. The irreversibility occurs because human societies, despite the advances of knowledge and technology, do not know how to reconstruct such complex systems. More so than in the scarcity of resources, it is in these planetary boundaries that we must seek the limits to our forms of growth. They are divided into two subsets.

1) For the ozone layer, the climate system, and ocean acidity, the boundaries can only be defined at a global level and with considerable uncertainty in scientific terms. In the case of climate change, the boundary is defined in terms of the accumulation of greenhouse gas emissions, which threaten the stability of the climate; in the case of the ozone layer, it is defined in terms of the emission of chlorinated compounds, which destroy the natural defense that the ozone in the stratosphere provides by filtering the sun's ultraviolet rays; and in the case of the oceans, it is defined in terms of the acidification of surface waters, through which the growth of living creatures at the lower end of the food chain (plankton, coral, shellfish, etc.) is seriously disrupted.

Among these first three boundaries defined at the global level, climate change occupies a central position because of interdependencies between regulatory systems: some fluoridated greenhouse gases also have a destructive effect on the ozone layer; the primary cause of ocean acidification is the increase in anthropogenic CO_2 emissions, which also leads to global warming. The three thresholds are thus not independent of one another. The fact that they can only be defined globally gives rise to a recurring problem for the implementation of corrective action, one that is well known in the world of international climate negotiations: individual actors get no advantage from taking unilateral action in the face of this type of environmental risk.

2) Six other boundaries have been defined by researchers at the Stockholm Resilience Centre, from which thresholds may be

calculated at different geographical scales.[5] Specifically, these concern the atmospheric concentration of aerosols—particles of either natural origin (e.g., clouds and dust from volcanic eruptions) or anthropogenic origin (e.g., compounds containing sulfur and dust escaping from boilers) present in the atmosphere; the maintenance of biodiversity, the loss of which can be estimated from the number of living species becoming extinct; the conservation of soil fertility; the freshwater cycle; the accumulation of chemicals; and the phosphate and nitrogen biochemical absorption cycles.

For each of these systems, the planetary boundary results from multiple thresholds at local or regional scales, as in the case of conventional pollution. But their aggregation points to a risk of irreversibility at a global level in response to the growth pattern of human activities. For example, the pollution of a river has consequences at the local level and represents a cost, sometimes difficult to reverse, for those living in the immediate vicinity. But these local pollution thresholds may be amplified or, conversely, reduced at the scale of the river's catchment area, which takes into account all the interdependencies affecting the regional hydrological system. Nor do the interdependencies stop there, because downstream from this basin, the discharges in turn affect the marine ecosystem into which the river flows. At the global level, it is therefore the entire water cycle of the planet that may be affected by human behavior. Among these many interdependencies, the boundary we shall be most concerned with here is the one constituted by the erosion of biodiversity. Virtually all forms of pollution may be seen as an assault on the reproduction of living species and therefore as adversely affecting the diversity of life. This is one of the issues in which the circular economy must go beyond the simple principle of recycling and begin to repair the existing damage. Indeed, thinking of the circular economy solely as recycling is to forget the initial harm done to the natural environment and consequently to ignore a large part of the problem.

Once natural capital is defined as a whole, covering on the one hand scarce resources present in their natural state and on the other the natural regulatory systems needed for their reproduction, the concept of the "green economy" becomes more intelligible: it is an economy that is based on these regulatory systems and views them as genuine factors of production in which it is necessary to invest. In such an approach, the green economy is no longer seen as a kind of return to a putative state of nature in which we have to rediscover our ancestors' ways of life, but as an increasingly complex system to be managed because levies on conventional natural resources undermine natural regulatory systems such as climate and biodiversity. As we will see, including these natural regulatory systems in the definition of natural capital radically changes the way in which the entire economy needs to work. While the classical rules of the functioning of the economy are perfectly able to give a scarcity value to natural resources by allotting an economic rent to them, they are still largely unable to assign a value to the large regulatory systems that constitute the framework of natural capital.

6

Hotelling

Beyond the Wall of Scarcity

HUMAN INGENUITY HAS, SINCE the beginnings of industrialization, constantly shifted the physical limits to growth, which hitherto had been seen as inviolable. With the digital revolution, some people were forecasting the advent of a dematerialized economy. Never before in history had so much capital been deployed to push back the wall constituted by the scarcity of raw materials: stepping up the race for energy-producing materials; the exploitation of unconventional hydrocarbons; purchase of agricultural land by foreign capital; cutthroat competition for access to rare earths, etc.

A powerful economic mechanism accounts for this accumulation of capital, namely rent, which can be defined as the additional value associated with physical scarcity. When it concerns scarcity of space, we speak of property rent: a square meter in Manhattan is worth tens of thousands of dollars, compared to just a few dollars in the plains of Nebraska and nothing at all in uninhabited desert regions. When it concerns scarcity of underground resources, we speak of mining rent: a barrel of oil from the oil terminals on the Red Sea sells for eighty to a hundred dollars, depending on price fluctuations, whereas its cost of extraction in the Saudi oilfield of Ghawar located only a few kilometers away is no more than two dollars.

David Ricardo was the first economist to foresee the role of rent in economic growth. The name of the American statistician and economist Harold Hotelling is associated in the economic literature particularly with exhaustible resources. His theory of exhaustible resources explains how the functioning of traditional markets pressures society to surmount successively announced walls of scarcity. However, these successive advances do not contribute to the preservation of natural capital as a natural system of regulation, as described in chapter 5. By helping society surmount the wall of scarcity, Hotelling exposes it to even greater perils. It is because of these dangers that our green capital approach leads to the introduction of the concept of "environmental rent," a new value that does not spontaneously arise in traditional markets, but whose function will be vital for the transition to a green economy.

How Long Until We Come Up Against the Wall? The Difference Between Reserves and Resources

When one retrospectively looks at the warnings that have alerted us in the past to the risk of an interruption of growth due to the exhaustion of natural resources, one often notices confusion between reserves and resources.[1]

The concept of resources refers to all deposits—speculative, presumed, probable, or proven—whether or not they have been discovered, whereas the concept of reserves is limited to those resources—probable or proven, evaluated or measured—that have been identified and are economically exploitable. Most of the time, when attempting to determine the time remaining for exploiting a resource, experts base their estimates on reserves, assessed at a particular moment in time according to given economic exploitation conditions and on the basis of established geological knowledge.

Calculations concerning the exhaustion of resources are generally based on the ratio between proven reserves—known resources that are recoverable with reasonable certainty and are economically exploitable given current prices and existing technology—and production or consumption. A more generous calculation involves taking into account an extrapolation of potential resources based on knowledge of geological formations and their connection with the resource. In both cases, for a given stock, simple division reveals the time left before it disappears.

A more sophisticated calculation, used in particular in the work directed by Meadows, involves taking account of a changing rate of demand over time. Typically, one needs to know the "initial" level of consumption—a reference level at a given time—of the resource as a physical unit and the total available reserves of the resource, and one needs to make an assumption as to the growth rate of consumption. It is then a matter of simply adding up all consumption to the point at which total consumption and the resource stock are equal.

These calculations ignore presumed resources (not yet discovered, but which are assumed can eventually be found at known and already prospected sites), speculative resources (undiscovered resources at not yet exploited sites, but where it is known how to find them), and as yet unsuspected resources. By changing the quantity of the resource available, this exclusion may have major effects on the evaluation of the time remaining for exploiting or consuming it. Twenty years ago, who would have imagined finding oil in sands in Canada or such large quantities of shale gas in the United States? Looking retrospectively at IEA figures, we see that the reserves of oil, gas, and even coal have increased over time.[2] Thus there may remain several decades with around forty years of oil available and with peak oil being reached every ten years . . . which is exactly what has happened since 1970.

Taking for the moment only the distinction between reserves and resources, we can safely say that the calculations pertaining to resource depletion may from time to time be worrying, but that is all. If, subsequently, we factor in technological developments with regard to exploitation, changes in the use of resources, economic conditions, and advances in geological knowledge, the wall of scarcity is probably not what we need to be most concerned about.

Economists took account of the physical limits of resources very early on. Beyond calculations of duration, the Earth is of finite size and the question of the limits to exploitation of resources is still relevant. One of the traditional contributions of economics is to show how this can be best achieved, namely by adding a "Hotelling rent" to the normal returns on capital invested in other respects in the economy. If conventional markets operate satisfactorily, the "invisible hand" will thus naturally prolong the periods of exploitation of the scarcest resources and thwart pessimistic prognoses, such as those of *The Limits to Growth*.

What's the Best Way to Hit the Wall?

The answer to this strange question may be found in the work of Hotelling, particularly his seminal paper "The Economics of Exhaustible Resources," published in 1931.[3] In it he presents the founding model for the optimal exploitation of an exhaustible resource. The question raised was the determination of the optimal rate of destruction, or the optimal use rate, of the resource.

The principle is very simple and still remains valid. An exhaustible resource stock is considered to be a specific asset producing revenue over time. The extraction and then the consumption of a resource unit means that it is impossible to extract and consume this unit at a later date because the stock—assumed to be finite and accurately known—is reduced as a result of this decision.

Therefore extracting a unit now entails a loss of future revenue. A firm seeking to maximize the present value of its returns is then confronted by this opportunity cost: giving up the value of the unit at a later date.

The extraction value of a unit extracted by the firm is equal to its selling price minus the cost of extraction. This extraction value is sometimes known in the literature as "scarcity rent" or "usage cost." If the firm decides not to extract, it is because the value of the resource in the ground, or the "in situ value," is greater than the extraction value. This "non-extraction value" represents the opportunity cost of resource depletion. At the margin, that is, for the last unit extracted, it must be equal to the extraction value.

It is very much as if you have to decide to sell an apartment now or in a year's time solely on the basis of a financial tradeoff. Selling now allows you to obtain the money immediately and put it into your savings account, while waiting a year implies that you expect the value of your apartment to grow faster than the return from the savings account. For this purely financial tradeoff to be a matter of complete indifference, the total return from the savings account must be equal to the expected price increase of your apartment. If you think your apartment will have a higher value, it is in your best interest to wait; if not, sell it today!

This same reasoning applies to the exploitation of a finite resource. For simplicity, we assume initially that the extraction cost of a resource unit is zero. The return from a financial asset and the market price per unit of the resource is then considered. By investing this amount in the financial market, an agent who owns this asset is assured of getting the total return from the asset as future income. Alternatively, he can extract a unit of the resource at a later point and sell it. By construction, under perfect competition, when all possible tradeoffs are implemented, the agent should be indifferent as to the two options:[4] 1) extracting a resource unit today and investing the proceeds from its sale in the financial market; and 2) extracting the resource unit in the following period and selling it. His indifference to the two alternatives implies that the returns are equal.

This is the logic that pertains in the "Hotelling rule" with zero extraction cost. It implies that at each period of time the value of the resource in the ground is equal to its extraction value. Dynamically, it requires that the growth rate of extraction value between different periods, and therefore the value in the ground, be equal to the opportunity cost of time, that is, the interest rate.

Considering now a non-zero extraction cost, we obtain the achievable margin, and therefore the possible investment in the financial market. The market price of the resource is then composed of two terms: its cost of extraction and its value in the ground. How the cost of extraction evolves depends on two opposite effects. Technical progress lowers the cost, but the resource becomes more difficult to extract as it is progressively depleted, which increases the cost.

Put schematically, at the beginning of the exploitation of an exhaustible resource the extraction cost is low, but it increases sharply when exhaustion approaches, irrespective of any technological advances. In the interim, it may increase or decrease, depending on which effect is stronger. Consequently, by lowering extraction costs, technological progress may lead to price changes following a U-shaped curve. Such a curve has been demonstrated by Margaret Slade for a number of minerals over the period from 1870 to 1978.[5] The difference between the value of the resource, which increases at the interest rate (and thus follows an exponential curve), and the evolving extraction cost constitutes the scarcity rent associated with the exploitation of an exhaustible resource.

We can then easily link this to the well known Ricardian rent associated with decreasing returns in agricultural production, as less productive land is brought into use. This rent represents the difference between the cost of production on fertile land and the cost of production on less fertile land. With the same quantity of work, the difference in productivity, and hence of the production of wealth between the two types of land, produces an income gap between the various landowners and farmers. Because the selling price of an agricultural commodity is set at the highest cost of production, a rent appears.

The optimal exploitation of a non-renewable resource in fact involves looking for the maximum amount of rents over time by determining at every moment the corresponding level of exploitation.[6] Under perfect competition, the optimal trajectory is reached when the Hotelling rent evolves in line with the interest rate.

Can We Go Over the Wall? Infinite Growth in a Finite World

The mechanism of Hotelling rent has historically been a powerful incentive for the discovery of new deposits, advances in operating technologies, and changes in economic conditions, making extraction operations profitable that were hitherto impossible. In addition there are changes in the use of resources that need to be taken into account.

The basic principle defined by Hotelling concerns the logic of raw materials extraction upstream of processing industries. Economists such as Partha Dasgupta and Geoffrey Heal, Robert Solow, and Joseph Stiglitz have also proposed taking into account the effects of technical progress or of substitution between physical capital and exhaustible resources that occurs throughout the usage chain of these resources.[7] By introducing a finite resource into the production function, they show that the limits to growth depend less on resource depletion than on the capacity either of technical progress or of substitution between factors of production to ensure growth. Furthermore, Hotelling rent spurs on the investment needed by freeing up increasing financial assets as the stock of resources is depleted.

This approach based on technical progress and confirmed by numerous historical observations can confound the pessimism of classical authors regarding potential long-term growth, and in particular simplistic projections from Ricardian rent. By gradually replacing land by physical capital (chemical inputs, mechanization, irrigation) and human capital (agronomy, genetics), agriculture seems to escape the physical limit of yields from the land.

In approaches that include technical progress, it is possible to obtain infinite growth even in an economy whose development is based on a finite resource. Indeed, if we agree that technical progress can constantly compensate for the depletion of the resource, then infinite growth is achievable. Intuitively, if it takes 10 liters of fuel to drive 100 kilometers and the gas tank holds 100 liters, then without technical progress the maximum distance from a tankful of gasoline is 1,000 kilometers. But by introducing technical progress, enabling the car to go 500 kilometers with 10 liters, then the maximum distance on a tankful becomes 5,000 kilometers. And if consumption performance is continuously improved, then it will become possible to drive any distance with the last drop of gasoline: advancing technology will have found perfect substitutes for continuing mobility without gasoline. Thus without too much difficulty we can imagine infinite growth in a finite world. Such thinking, however, is reminiscent of Zeno's paradox, in which the arrow never reaches its target. It is a question here not of failing to grasp temporal dynamics, but of a rash belief in technological progress.

Up until now, historical analysis seems to have vindicated those who believe they can jump over the wall of scarcity. If indeed this is possible, what then will be found on the other side? The answer is, in all probability, natural capital that is even more damaged, as becomes evident by introducing the effects of pollution and the saturation of the ecosystem, in short by including what economists call "negative externalities."[8]

Might It Not Be Better Beyond the Wall?

Paradoxically, it is probably not the scarcity of exhaustible resources that is the main limit to economic growth, but the conditions of reproduction of all the planet's resources. This reproduction is threatened by the accumulation of pollution resulting from the ceaseless search for exhaustible resources.

According to the diagnosis by the European Commission, mankind has altered the planet's ecosystems more rapidly and more extensively over the past fifty years than during any other period in human history. This degradation has adversely affected a number of environmental services and these in turn hamper the conditions for growth. It may firstly be a matter of production services such as food, water, pharmaceutical goods, and genetic resources, and here it is possible to calculate the production shortfall for the activity concerned. But it may also involve environmental regulation services such as air quality, soil erosion, climate, or the cushioning of the effects of natural hazards (tornadoes, tsunamis, etc.), where it is more the quality of growth that is adversely affected.[9] Each of these services is negatively impacted through pollution resulting from exploitation, production, and consumption activities.

The *Millennium Assessment Report* has shown that the main ecological services provided by the biosphere are now under pressure.[10] With regard to biodiversity, which can be detailed at a local level, the report has also highlighted the direct links between investment in rebuilding an ecosystem and the economic benefits stemming from it. For example, the restoration of degraded mangrove forests in an African delta can generate multiple benefits for local populations: the rebuilding of fisheries, protection against floods, protection against soil salinization, and opportunities for new crops.

The examples of energy and climate also provide good illustrations of the process of altering a natural regulatory system.

Between 1973 and 2013, the geographic redistribution of growth speeded up CO_2 emissions, the accumulation of which in the atmosphere toward the end of this period has reached thresholds considered by the scientific community to be high risk. Contrary to a persistent idea, the increase in known oil reserves following their apparent scarcity has not led to a slowdown in greenhouse gas emissions related to energy use. The efficiency gains generated in the short term have been offset in the medium term by shifts to other fossil fuels (gas and especially coal since 2000) and by the incentive to recover a larger proportion of the stock of oil underground. Yet, as the scientific community is making clear with its increasingly insistent messages, the use for energy purposes of the carbon stock available underground would, in the context of current technology, constitute a major threat to the stability of the climate. Exploiting fossil resources beyond a certain threshold impairs the atmosphere's capacity to ensure climate stability, which—along with biodiversity—is one of the two major regulation systems of natural capital.

From Hotelling Rent to Environmental Rent

As long as natural capital is viewed as the sum of scarce resources, its introduction into the economic system occurs spontaneously through the introduction of Hotelling rents into traditional markets for goods and services. But once this capital is included in the regulation systems underlying the reproduction of resources, its introduction into the economy raises new questions: Is it necessary to determine a price associated with the use and reproduction of these natural regulation systems? Let us call this price "environmental rent" because its purpose is to pay for the investments needed to protect natural capital. How is it to be calculated in theory and in practice?

Generations of environmental economists have attempted to determine this price on the basis of the concept of "externalities" and introduce it into the calculations of economic agents. The following chapters show the benefits of this approach and will extend it with a representation of the economy and of growth in which natural capital is viewed as a new factor of production. For the moment, we can give an illustration of it by considering climate change, the area in which research is probably most advanced.

As we have seen, our economic system already links Hotelling rent to the exploitation of hydrocarbon resources, commonly referred to as oil rent. This rent can be estimated at about three trillion dollars in 2013, and this is added to the "normal" returns on capital used in the economy, known by economists as the average profit rate. In fact, most of this rent is captured by taxation, on both the production side and the consumption side.

Oil rent, which increases over time in line with (constantly deferred) projections of peak oil, acts as an incentive for investment in exploration and research. Whenever these investments lead to new discoveries, a temporary abundance of fossil energy occurs, thereby interrupting the secular increase in the rent. This occurred with the oil countershock of the 1990s and more recently with the exploitation of shale gas deposits, which has led to the quadrupling of International Energy Agency estimates of global gas reserves. If we represent natural capital through the simple aggregation of non-renewable resources, as in economists' standard practice, there is every reason to be delighted that the wall of raw materials scarcity has again been pushed back in time.

Let us now introduce the climate's regulatory function, an essential component of natural capital that regulates the planet's incoming and outgoing energy flows through the greenhouse effect. According to IPCC figures, the amount of carbon tied up in underground resources (not reserves) in the form of fossil fuel deposits is about four times the amount present in the atmosphere as CO_2. With regard to the climate, the major concern is therefore not the shortage of fossil fuels but the excess of carbon at risk of being released into the atmosphere by continuing to extract fossil fuels. But this systemic climate change risk is not taken into account by Hotelling rents, which traditionally price the scarcity of natural resources, thereby compelling economic agents to constantly push back the wall of scarcity. In concrete terms, this means that economic decisions do not at present incorporate the value associated with the climate's regulatory function.

How does one define the economic value associated with climate protection? As a first approach, it can be defined as the price or cost that society will agree to pay to avoid emitting a certain amount of greenhouse gases into the atmosphere. This price is more commonly known as the carbon price. Various countries and some U.S. states have begun to introduce it in the economy following the Kyoto Protocol: the European Union, Japan, California, the northeastern states in the United States, and now China and South Korea.[11] Suppose that all countries of the world agree to set a price of \$25 per ton of CO_2 equivalent. Global greenhouse gas emissions in 2013 have been estimated at about fifty billion tons of CO_2 equivalent.[12] At a price of \$25 per ton, this amounts to a worldwide total of about \$1,250 billion. This value is a new rent: the environmental rent associated with the scarcity of the atmosphere with regard to its climate regulation function. But this \$1,250 billion represents only half the estimated value of oil rents for 2013. Furthermore, it only concerns pricing the climate, just one of the nine major regulatory functions of natural capital identified by the Stockholm Resilience Centre. It is evident, therefore, that it will take a long time for the economy to escape the Hotelling paradigm. Although the present economic system is clearly able to price stocks of materials through Hotelling rents, the valuation of natural capital through environmental rent is still in its infancy. As everyone knows, you cannot put a price on nature!

7

Nature Has No Price

*How Then Is the Cost of Its
Degradation to Be Measured?*

"WHATEVER THE PRICE GIVEN to the *Mona Lisa*, that would say nothing about its value."[1] This aphorism by the French economist and anthropologist Jacques Weber expresses the perplexity experienced by economists faced with the problem of valuing natural capital. Comparison with a work of art, which in principle is unique and irreproducible, is very relevant: in the reproduction market, thousands of copies of the *Mona Lisa* are sold every year in the Louvre Museum shop (and elsewhere) at prices that vary according to the quality of the reproduction, its size, storage costs, etc. These copies are just like any other merchandise. But the *Mona Lisa* itself has no price: its value is inestimable, like that of Notre Dame Cathedral, the works of Victor Hugo or William Shakespeare, and . . . nature. As well as being non-reproducible, nature is a public good, that is, a good that is held in common. It is also indivisible. No more than one can imagine cutting up the *Mona Lisa* into small squares like carpet samples and selling them, can one slice up nature and trade its components in a market. There is no question here of "commodifying" nature, as the expression goes.

While it makes no sense to put a price on nature, there are nevertheless many situations in which society must evaluate the costs of its use or its destruction. Introducing the idea of an environmental good into decision making should rightly reflect the use made of nature. In this sense, nature can indeed be valued and should be taken into account in economic decisions. In March 1989, the oil tanker *Exxon Valdez* went aground in the Strait of Prince William in Alaska and spilled 42,000 tons of crude oil into the Arctic Ocean, producing an oil slick that seriously damaged a very fragile ecosystem. Five years after the disaster, Exxon Mobil was fined $5 billion. After an appeal, in 2008 the U.S. Supreme Court reduced the amount payable for damages and interest to $500 million. In the end, Exxon spent nearly $3.4 billion cleaning up the coasts and wildlife and on compensation for fisheries.

Why $500 million rather than $5 billion? And why not $50 billion, or nothing at all? In both cases, the judge's decision was preceded by an expert assessment of the damage, the value of nature, and the costs of its degradation. But it is clear that we are far from knowing how to price environmental damage because there is basically no consensus on the methods used for assigning a value to nature and establishing a cost for its use and a value for its protection. Let us now see what economists can and cannot do in this regard.

From the Enchanted World of General Equilibrium to the "Tragedy of the Commons"

In the enchanted world of economic theory, prices play a central part. They reflect use values and enable assessment of the gains or losses in welfare that consumption or production of a good brings to an individual or to the community. If I decide to buy a "baguette tradition" (baked without any additives, unlike an

ordinary baguette), at €1.15 instead of an ordinary baguette at
€0.90, it is not simply that my budget allows me to do so, but
also that the additional pleasure derived from eating this particular type of baguette is worth at least €0.25. If it is worth more, I
realize a surplus, and if it is worth less, I go back to the ordinary
baguette. My baker sets the price of the baguette on the basis
of his costs, market conditions, and what he believes I am willing to pay. Agents ultimately base their economic calculations
on the price system. Economic theory provides two fundamental
theorems of welfare to show under what conditions society can
maximize its welfare.[2] Here we use only the first of these, leaving
the other until chapter 17.

The first welfare theorem asserts that if all agents behave com-
petitively in markets in which there is perfect information, then
market equilibrium leads to a situation in which it is impossible
to improve the situation of one or more agents without worsen-
ing at least one other agent's situation. This situation is known as
the Pareto optimum, named after the celebrated Italian economist
Vilfredo Pareto. The Pareto optimum is thus a criterion of efficiency
in the use of resources in the economy. It should be noted that this
criterion has nothing to say about fairness and social justice.[3]

The conditions for achieving such efficiency in the allocation of
resources are highly specific and restrictive, making it difficult to
transpose the theorem into real life. Information must be perfect
(we know everything about everything), people do not engage in
strategic behavior (there is perfect competition), all goods have
a price (markets are all-encompassing), and there are no public
goods (absence of collective goods). Nevertheless, the theorem
does highlight one crucial element: people's decisions are based
on a tradeoff or choice among different values reflected by prices.
The more the various values are known, the better the decision making. By increasing the capacity to measure the value of
resources, there is a gain in economic efficiency if the markets
function properly.

But what happens if a particular good, say an environmental good, remains unpriced? The answer is found in the title of the influential paper by Garrett Hardin published in 1968, "The Tragedy of the Commons," which permanently influenced environmental thinking in the United States.[4] In this paper, which is in fact more concerned with demography than the environment, Hardin brilliantly shows how an environmental good being free leads to its overuse, or even destruction, in a society dominated by market exchange. For his demonstration, he draws on the example of a freely accessible communal pasture that, in a feudal village society, serves as social security. Every villager, whatever his status, is free to take his livestock to graze in the pasture. Though effective in an unchanging economy, this type of organization does not support growth: increasing the number of animals leads to overgrazing of the meadow. Because its degradation imposes no costs on any of the villagers, it is in each herdsman's interest to use the pasture for as long as any grass remains. The increase in the number of animals leads to the destruction of the environmental good. Historically, the "enclosures" movement, beginning in England in the sixteenth century, involved distributing ownership rights on such land, which then acquired a market value. The owners of the land consequently hastened to fence it off so as to protect it from encroachment by livestock from the village. Hardin reminds us that the issue of measuring the value of environmental goods is not limited to cases of compensation, as in the case of the *Exxon Valdez*. In the absence of valuation, there is a great risk of overexploiting nature, or even destroying it. If its value is zero, why pay any attention to it? Nobody suffers the costs and nobody assesses the consequences of his actions. No one will take this zero value—we forgot to assign it a price!—into account in his consumption or production decisions or his behavior. One of the tragedies of environmental goods is that generally they are not commodities and therefore remain unpriced. It may be regrettable that a monetary value has to be put on something for it to be taken into account. But until such time as the economy is regulated by mechanisms other than the market, an answer must be found if the "tragedy" of the destruction of environmental goods is to be avoided. The chapters that follow will examine the various options for implementing such pricing. Before moving on to practical considerations, it is worthwhile to examine the theoretical underpinnings of the valuation of natural capital.

Anthropocentrism, Biocentrism, or Ecocentrism?

Whether it is a matter of a resource, pollution, or, more generally, an externality, it is not easy to assess the different possible values of an environmental good. A forest may be used for the exploitation of timber but also for walking in today, tomorrow, or in ten years' time. In addition, it produces and reproduces biodiversity and plays a role in water regulation and in the carbon cycle. A lake can be used for power generation, recreation, irrigation, biodiversity, and so on. Economists generally consider that an environmental good has a total economic value (TEV) that can be broken down into three components:

- a direct use value (the wood I exploit in my forest or the welfare I obtain from walking in it);
- a non-use or indirect use value (the wood I leave to my descendants or the reserve of biodiversity in my forest that contributes to the collective heritage) and
- an intrinsic value that exists independently of mankind (the value of the species living there and the ecosystems they form).

Each of these components may be linked to philosophical or moral foundations.

In the Cartesian tradition, the individual is at the center of things and, thanks to his knowledge, becomes master of circumambient nature. Mathematics and physics are, as much as philosophy, the disciplines of Descartes, and consign to the scrapheap the physics of Aristotle and the animism of the Middle Ages. Nature becomes a kind of reservoir of materials and resources from which mankind can draw as much as is needed. This narrow anthropocentric viewpoint draws on use values alone for evaluating nature, which is seen as having no value other than what it provides directly to mankind. If the value of nature is limited to its direct uses, we remain in a wholly utilitarian, Benthamite perspective.

Spinoza, a contemporary of Descartes, developed a completely different conception of the individual and of nature. In *Ethics* (1677), his major work, the individual is indissociable from a more complex whole, just one of the many "attributes" of "substance" that encompasses him and without which he would not exist. In contrast with Cartesian individualism based on the *cogito* is the Spinozan view of *conatus*, according to which a person's being persists by virtue of entering into harmony with the whole. In this conception, the value of nature is no longer reduced to the uses that may be made of it but rests on the complex system of interdependencies that links the whole to its parts, substance to its attributes, and nature to mankind. Equally, in Kant's *Foundations of the Metaphysics of Morals* (1785), the "categorical imperative" provides a moral grounding for this non-use value. Indeed, Kant recommends regulating day-to-day life in accordance with precepts that incorporate his desired goal of universal concord.

These Spinozan and Kantian approaches considerably broaden the utilitarian view of nature but are nevertheless still anthropocentric in the sense that the value of nature continues to be determined in relation to human beings. In contrast, the biocentric and ecocentric approaches consider that nature has an intrinsic value, linked to living creatures in the former and to ecosystems in the latter. In the biocentric approach, living creatures have a value independent of mankind, and there is, for example, no reason why an animal should give up its life on behalf of a human being. Biocentrism thus advocates "biotic egalitarianism." Ecocentrism, in contrast to biocentrism, considers that the basis of value is not individual but collective. Value does not reside in living creatures themselves, considered individually, but in complex wholes in which they find a place—in other words, ecosystems or, better yet, "the biosphere." Aldo Leopold succinctly expresses the golden rule of ecocentrism: "A thing is right when it tends to preserve the integrity, stability and beauty of the biotic community. It is wrong when it tends otherwise."[5]

As can be seen, the philosophical foundations of environmental values may lead to unorthodox approaches because they question the reasons why society protects nature or overexploits it for profit. More fundamentally, they come down to the prevailing conception of the world. As economists, we are inclined to adopt the anthropocentric approach in which the value of nature is determined in relation to mankind and its preferences. But as we shall see, the green economy is not compatible with the narrow deployment of use value, in keeping with the Cartesian mechanism (nature's "principle of inertia"). It needs to be enriched with broader notions, whose philosophical roots may be found in Spinoza, Descartes's great opponent in the seventeenth century, and in Kant's moral justification. Let us now attempt to transpose these philosophical ideas to the economist's toolbox.

The Value of Nature in an Enlarged Anthropocentric View

Use values represent the proportion of total economic value associated with the use of a good, for example, the value of a forest path when one walks along it or the value of exploiting the forest's timber. Determining these values must also take time into account: the forest's use value depends not only on the services it can provide today, but also on the value placed on being able to walk in the forest in ten years' time or the value of the timber thirty years from now. This dimension of deferred use in time allows us to introduce the concept of "option value."

Retaining the option of using or not using the good at some later point is an integral part of its use value. The concept of option value, briefly introduced by the American economist Burton Weisbrod in 1964,[6] refers to the benefit that may arise from conserving the good over time. Correspondingly, it adds to the cost of the destruction of that good. By definition, we are never certain of the future and thus we are not even sure we will still want to walk in the forest in ten, fifteen, or twenty years' time. But keeping this possibility open today nevertheless has value.

Another type of option value, emphasized by Kenneth Arrow, Anthony Fisher, and Claude Henry, touches on notions of information and irreversibility.[7] The idea is that the arrival of information over time has value and that faced with a decision having irreversible or near-irreversible consequences, such as the construction of a dam, a value should be attached to this irreversibility. In the case of the dam, a valley is flooded and all the life in it is destroyed. In making the decision to build the dam, it is important to be sure that it will function as intended. Taking into account the irreversible aspect, the return must compensate for the fact that the valley can no longer be used for anything else. The discussions launched in 1919 on the possibility of constructing the Three Gorges Dam in Hubei province in China emphasized the fact of its irreversibility. At an estimated cost of forty to fifty billion euros and with the displacement of more than 1.3 million residents, the project was finally adopted on April 3, 1992, on the strength, unusual for Communist China, of only a two-thirds vote in favor.

Non-use values may first be defined as option values concerning a future utility and the margins of freedom retained against the risk of irreversibility. They also concern use values for others, be they relatives (such as one's descendants) or complete strangers.

We then speak either of bequest values or altruism. These values, commonly used in environmental economics, clearly lie within the tradition of Kantian philosophy.

A second category of non-use values concerns the regulatory functions of natural capital and the risk of their breakdown, both locally and globally. At the local level, a classic example of such a breakdown of regulatory functions is the collapse of an overexploited ecosystem, whose regenerative capacity it is very hard to know in advance. If an ex ante value is not given to these regulatory functions, which support multiple services that are used only very partially, there is a serious risk of paying dearly ex post when the system collapses. A well known example is the failure of the fishing moratorium imposed by the Canadian government to bring cod back to the waters of Newfoundland and Labrador. The explanation is that over the years fishing practices altered the regulatory capacity of the marine ecosystem by destroying the habitats where the species reproduced. Not assigning a value to this ecosystem thus generated quite unanticipated costs, in that the ending of catches has not enabled the stock to replenish itself.

The example of overfishing may be transposed to the macroeconomic level. In chapter 5, drawing on the concept of "planetary boundaries," we identified the major regulatory functions constituting natural capital. Economic growth depends on increasing use of natural capital, a factor of production without charge. Because of this free availability, the growth of the economy progressively alters certain regulatory functions, especially climate and biodiversity. This mechanism is not properly represented in standard growth theories. It is time to introduce green capital into the modeling of growth.

8

Beyond Hotelling

Natural Capital as a Factor Required for Growth

GREEN GROWTH IS GENERALLY presented as a form of expansion that is an alternative to degrowth but which would entail the continuation of traditional models of resource predation. Yet in the absence of a serious analytical basis, this approach merely describes an outcome without providing any indication as to the conditions required to attain it. As a result, it becomes possible to imagine as many paths to green growth as there are forms of non-green growth, and sometimes even to disguise, rename, and confuse them. The ministry of the environment will identify its "green jobs." Experts will calculate that there is 25 percent "green investment" in a given recovery plan. Hypermarkets will flaunt their latest "green" recyclable bags. In the absence of a rigorous definition, "green" is everywhere and nowhere. It becomes a kind of variable geometry that everyone can adapt as required without having any comprehensive totalizing view of it.

The concept of growth conventionally deployed by economists covers all the mechanisms leading to an increase in wealth that can be understood in terms of production, income, or living conditions. In itself, such growth is in principle neither "right" nor "wrong." To characterize it, it is necessary to examine the mechanisms related to the factors mobilized, the remuneration of these factors, and the wealth produced. Its description depends on how natural capital is treated in the growth process. Growth should be termed "green" only when it includes natural capital among the factors of production in which it is necessary to invest, in the same way as labor or capital.

Conceptually, this definition is straightforward enough. But how is it to be used in practice if it is impossible to put a price on nature, that is to say, to directly assess the value represented by the various services provided by the stock of natural capital? We showed in chapter 7 that it is relevant at the micro level to estimate the cost of damage to the environment caused by economic activities. It is on the basis of such methods that we shall construct our theoretical representation of the macroeconomic production function, thereby gaining a better understanding of green growth. Thus it is not the stock of natural capital directly that we use in our approach, as suggested by common sense, but its degradation by pollution.

Depicted in this way, it will be seen that our view of growth allows us to better formulate the questions concerned with green growth. In particular, our approach shows that by assigning the right prices to these forms of pollution, the condition of natural capital as a factor of production is improved. We should then understand natural capital either as a source of growth, once it can be renewed and made to flourish, or as an inescapable brake on expansion, if it is allowed to deteriorate.

Sources of Growth in the Standard Representation of Production

The standard representations of economic theory analyze production as a combination of human capital and physical capital, with pollution as a possibly inevitable side product entailing the deterioration of natural capital and its regulatory functions. From the standpoint of standard growth theory, economists are generally and systematically in agreement on the driving role of accumulation, both of human capital and physical capital.

Human capital represents work in terms of contribution to production. This concept encompasses both demographic variables influencing the size of the workforce (birth rates, longevity, immigration) and aspects of labor productivity in terms of qualifications and educational levels. Adding equally qualified employees or replacing them by fewer, more qualified workers amounts, in terms of their production efficiency, to the same thing from strictly a production standpoint. In theories of growth with human capital, it is essential to augment the labor force, either quantitatively or qualitatively. In the latter case, it is the accumulation, in principle limitless, of knowledge that is crucial to growth. By raising the educational level, the overall production capacity of the economy is improved. This is one of the myths that is often willingly ascribed to the digital economy and that too quickly forgets the necessary production of real goods accompanying this growth.

The second element concerns the increase of physical (or productive) capital. Productive capital covers all investment, machinery, tools, infrastructure, and inputs other than the labor involved in the production of wealth. An increase in productive capital thus requires an investment effort that allows the stock and performance of the capital mobilized in the economy to be increased. For growth, as for the workforce, the important thing is therefore the increase of physical capital, quantitatively or qualitatively.

Put simply, economists represent the combination of these two factors of production (capital and labor) by means of a production function that sets out all possible combinations of the factors and accounts for the level of output of an economy. On this basis, investment, education, innovation, and technical progress enable the productivity of factors to rise, either individually or for the production function as a whole. In the latter case, a multiplication factor that positively impacts all production is introduced.[1]

We can consider, as in conventional growth models, the existence of positive externalities taken into account in this multiplying factor. This factor variously incorporates technical progress, the average level of human capital, and the average level of physical capital. We then obtain a production function with increasing returns and potentially infinite economic growth. In such systems, only the capacity to increase production factors counts, that is, the accumulation of physical capital, the accumulation of human capital, or the multiplying factor, without considering their possible environmental effects.

A great many growth models have thus been conceived, among them an exogenous growth model linked to ex nihilo technical progress like that from Robert Solow (in 1956) as well as endogenous growth models linked variously to human capital (from Robert Lucas in 1988 and Sergio Rebelo in 1991), physical capital (from Paul Romer in 1990), and public investment in education or research and development (from Robert Barro in 1996).[2] All of these approaches ultimately raise the question of the constitutive elements of the production function, the possibilities of taking into account their effects on growth, and ways of deploying them so as to encourage their accumulation.[3]

As we saw in chapter 5, in response to the Club of Rome, authors such as Dasgupta, Heal, Solow, and Stiglitz have proposed taking into account the effects of technical progress and of substitution between physical capital and exhaustible resources. By introducing an exhaustible resource into the production function, they show that the limits to growth are less about resource depletion than about the capacity of technological progress to ensure continued growth. These models are based on the assumption of possible substitutions between factors, an assumption that is plausible enough for stocks of non-renewable resources but much more questionable with regard to natural regulatory systems such as ecosystem diversity or the climate.

Yet there is much more to fear from damage to the reproduction conditions of resources, disruption of climate control systems, and the loss of biodiversity than there is from the potential exhaustion of resources.

Taking Account of Pollution in the Production Function

To provide a framework for our analysis, let us consider the case, alluded to in the introduction, of a shepherd making a living by producing wool from sheep shearing and washing the raw wool. Suppose the shepherd is relatively efficient at small-scale shearing, with ten shearings and five fleeces an hour. Because washing the wool emits pollutants into the used water, the owner is no longer able to take water from the stream as he used to and therefore asks the shepherd to shear the sheep and wash the fleeces without using water. As this is more difficult, the shepherd still manages to shear ten sheep but can only wash two fleeces an hour. Hence the productivity of the water resource corresponds to the three missing fleeces. Part of the creation of value is thus attributable to the use made of the water.

Recent work on the concept of growth, especially the approach adopted by Acemoğlu and Dasgupta,[4] emphasizes the importance of taking account of natural capital in economic development. Thus when the production of a good results in pollution, then with the production technique unchanged, a reduction in pollution is accompanied by a decrease in production. Because this all takes place as if pollution were a constitutive element of production, it is therefore essential that pollution be integrated into the production function. The approaches proposed in particular by Brian Copeland and Scott Taylor (in 1994), Nancy Stokey (in 1998), and Pierre-André Jouvet, Philippe Michel, and Gilles

Rotillon (in 2005)[5] precisely allow this inclusion of pollution as a factor of production.[6]

Intuitively, if we consider a production function with the usual factors—physical capital and human capital—and if this production generates pollution, then we can define an intensity index of pollution that depends on the degree of use of the technology. This index gives the relationship between the level of production and the level of pollution. Therefore we can establish a link between the degree of use of the production technology and the intensity of pollution. The latter should not be seen as a mere externality or side product but as a factor of production in its own right. The production function thus increases from two to three factors—physical capital, human capital, and pollution, the last of which represents the natural capital used in production. If we no longer view pollution as a simple consequence of production activity, it is because there is a complementarity between pollution and the other factors of production.

This formulation may at first sight seem diametrically opposed to the standard view of green growth as balancing economic progress and the reduction of pollution. It nevertheless corresponds well to reality, even though economic actors will conveniently ignore it as long as the act of polluting is free. Once pollution has a price, however, it is immediately taken into account. Since the introduction of the European system of CO_2 allowances, all the major electricity companies have fully accepted that a ton of CO_2 has become a factor of production whose sourcing they must ensure, along with coal, gas, or any other fuel, to keep their power plants operating.

The main lesson from this new approach to supply is that part of the production associated with traditional factors of production ultimately depends on pollution. This is not simply a matter of nomenclature, because as soon as we consider three factors instead of two at a given level of output, we necessarily see a decrease in the respective weights of physical capital and human capital in production, a decrease that is offset by taking into account the environment. The main consequence is that the productivities of the usual two factors need to be revised downward. Indeed, in accepting that the environment is involved in production, the productivities of physical and human capital are in fact overvalued for a given level of production. The productivity of these factors, and therefore their remuneration, should be corrected because part of the productivity attributed to conventional factors of production ultimately stems from the degradation of the environment.[7]

Up to this point, we have limited ourselves to describing the mechanism of complementarity between pollution and production in an economy in which actors have no incentive to reduce their damage to the environment. We have not included the impact of pollution on the overall productivity of factors. We can interpret this impact as corresponding to a certain state of nature that will affect the supply capacity of the economy.

To go a little further in understanding the growth process, it is necessary to include the feedback effect of pollution on the overall productivity of factors. Initially, we need to introduce a negative feedback effect showing that although the increase in pollution contributes to short-term growth, in the medium and long term increased pollution destroys the sustainability of growth. The introduction of pricing for pollution creates a new value in the economy that we term "environmental rent." This rent leads actors to combat environmental damage and consequently to act positively on the productivity of the economy in the long term. Natural capital then becomes a genuine source of growth.

Natural Capital as a Source of Growth: Toward the Restoration of Regulatory Functions

Like productive capital and human capital, natural capital covers a very broad spectrum, ranging from exhaustible and renewable resources to all the services provided by the environment. Those aspects of natural capital pertaining to exhaustible resources are not difficult to understand when it is a matter of resources such as oil, coal, gas, or uranium. It should be noted that the use of these resources usually involves fixed physical capital. Increasing investment in the extraction of exhaustible resources, as the stock is progressively depleted, is financed by a Hotelling rent, which also increases over time. This rent is incorporated into the return on the physical capital.

In the case of renewable resources such as forests, agricultural resources, fisheries, water, and biodiversity in general (defined in terms of the number of genomes), things are more complicated. The economy can work only if there are natural, self-reproducing regulatory services. Note also that these services are clearly indispensable for the reproduction of human capital, whereas physical capital is more resilient to pollution and, more generally, to the degradation of natural regulatory systems. The deterioration of these services may lead to the depletion or even disappearance of goods previously viewed as "free," that is, available in unlimited quantities, such water, air, climate stability, biodiversity, and so on.

When we look at the various services provided by nature, the spectrum tends to enlarge to production services, regulatory services, primary services, and cultural services. Production services include food, energy resources, water production, pharmaceutical resources, genetic resources, and aesthetic resources. In terms of regulatory services, there is air quality, erosion, water purification, climate regulation, and the cushioning of the effects of natural hazards such as tornadoes and tsunamis. Primary services comprise soil formation, photosynthesis, the water cycle, and the cycle of nutrients essential to life. Finally, it is important not to overlook all the services involved in cultural creativity, inspiration, and educational values, as well as a host of recreational activities ranging from the contemplation of a landscape to outdoor sports activities. Retroactively, the deterioration of these different services may negatively impact, directly or indirectly, the planet's production capacities. The emergence of a number of diseases associated, for example, with air pollution can degrade human capital, thereby generating a significant cost to the overall economy, along similar lines to the degradation of the natural elements constitutive of productive capital, such as the increasing scarcity of timber, the loss of biodiversity, and the deterioration of groundwater.

It is thus clear that natural capital plays a significant role in the productivity of the two production factors generally considered. Consequently, it is reasonable to assert that part of the total creation of value is directly attributable to environmental capital and should be included in the factor measuring overall productivity.

Exogenous and endogenous growth models ultimately ascribe a key role to the overall multiplicative parameter of the production function and attempt to put it into perspective. If we agree that the conditions of reproduction of human capital and physical capital partly depend on the state of the environment, then the multiplicative factor of the production function must depend on the overall quality of the environment. Thus the production function taking account of the environment depends not only on the pollution emitted during the production process, but also on the accumulation of this pollution, thereby changing the conditions in which production is carried out.

Viewing the quality of the environment as a decreasing function of the stock of pollution, it is apparent that the flow of pollution emitted increases the stock of pollution and thus degrades the conditions of production itself. The quality of the environment becomes a constitutive aspect of growth, and this necessitates developing a model that can consider pollution both as a factor of production and as a cause of the limitation of growth.

As soon as we introduce a price or a constraint to take environmental externalities into account, it creates an incentive to reduce these externalities (pollution). This reduction of externalities can lead to a gradual improvement in production's overall productivity factor. It then becomes a question of who will be willing to pay what we have termed "environmental rent." To take its place in the economy, this new rent should be levied on the factors producing environmental deterioration, in other words, on preexisting profits through the return on capital and/or wages.

In practice, setting such prices to protect natural capital does not depend solely on the economic calculations made by informed planners. More than anything it depends on the power relations among the economy's various actors, none of whom want to erode their earnings in the interests of natural capital. The incorporation of natural capital in the production function does not occur gratis. So who will agree to pay the environmental rent needed to finance the reproduction of natural capital?

Natural Capital and the Remuneration of Factors of Production

One of the basic principles of economics is that every factor of production receives remuneration proportional to the wealth created by virtue of its contribution to total wealth, that is, a "fair share" of the creation of value. Because each factor of production can be broken down in greater or lesser detail, no one and nothing is in principle forgotten: human capital ranging from the employee without any recognized qualification to the senior executive with a whole string of degrees, and productive capital from the handheld hammer to the nuclear power plant. In this simplified world, employees receive the share from human capital and the various owners of productive capital receive its share in the creation of value. Everyone consumes and saves according to their income and their preferences among different goods. This economic theory works perfectly well and the increases in the various factors of production constitute the source of growth. All that remains to be agreed on is therefore the true share of each factor in the creation of wealth.

The introduction of natural capital into the production system fundamentally changes the remuneration of traditional factors of production and their combination in the production function. In addition to wages and profits, which allow the reproduction of labor and capital, the remuneration of the environment is added as the third factor of production.

This addition entails reviewing the distribution of revenues within the economy, and more specifically within the production system. We must therefore revisit the distribution of wealth among human capital, productive capital, and natural capital. "Fair" compensation for natural capital needs to be implemented through a levy on existing revenues. If capital is the source of pollution, for example through the use of coal in a power plant, it seems appropriate and logical that the levy be made on the return on capital. If it is the combination of labor and capital that is the source of pollution, then the levy should be made in proportion to the contribution of each factor in the production process.

In both cases, the initial distribution of wealth between factors of production is changed because a share of the productivity that is in reality attributable to natural capital is ascribed to them. This remuneration of natural capital by virtue of its contribution to the production of wealth offers the possibility of an investment in its reproduction and thus in the improvement of the conditions of growth. By introducing the degradation of natural capital into the production function, the question arises as to who should foot the bill. Conversely, investing in the quality of the environment may lead to an increase in overall production. Who then will be the beneficiaries? Such are the key issues that allow us to devise green trajectories with very different social and political implications.

This approach should make us think not only about the distribution of wealth among the various factors of production, but also about the very conditions of growth. Investing in environmental quality implies a diminishing role for pollution in production, as well as an improvement in all the conditions of production and thus the conditions for growth in the medium term. In practice, does this lead to more growth or less growth?

The Introduction of the Pricing of Pollution
and Its Impact on Growth

The introduction of efficient pricing of pollution provides a way of propelling investment toward an economy that is both low carbon and more environmentally responsible, as well as encouraging direct investment in natural capital. There are therefore two levers through which changed behavior will impact the growth path: the fight against pollution, which is activated by its pricing, and direct investment in the quality of the environment. In our model, the fight against pollution has mixed effects on growth—negative in the short term, because one factor of production is reduced unless there is substitution toward the other two factors, and positive in the medium and long term through the positive feedback effect on the quality of the environment factor. Direct investment in the quality of the environment may result in incentives giving a value to environmental services, for example, when a value is placed on biodiversity and its protection such that it is no longer deemed sufficient simply to restrict access to natural reserves.

If we take the three-factor function, we need to explain the determinants of the multiplicative factor of production. We know that a healthier environment allows better reproduction of physical capital and increased productivity of labor, but we also know that biodiversity represents a potential for growth. The multiplicative factor of the production function is directly related to the state of the environment at all times. Formally, given that the flow of pollution in each period increases the stock of pollution, there is consequently an overall reduction in the quality of the environment and a limitation of growth potential. Thus investing in the quality of the environment and in the repair of its regulatory functions increases this factor, and therefore not only maintains growth potential but also improves it. A pure and plentiful water supply necessarily helps the shepherd produce fleeces.

Leaving aside the long litany of green jobs, we can focus on an essential element, namely investment, the key factor in the ecological transition. If we take the consolidated view of productive capital in the sense of genuine savings (chapter 4), such investment should be seen as an early substitution between conventional physical capital and natural capital. The pharmaceutical industry, along with the beauty product and food industries, has clearly perceived that investing in biodiversity is in its interest.

As we have seen, the emerging countries are potentially new markets for renewable energy and low-carbon technologies. They are waiting for new technologies to enable them to meet the growing needs of their people, which are now engines of growth. Pricing pollution and the investment in natural capital that doing so makes possible will play an increasing role in international competition in the future. The impressive ramp-up of China with regard to renewable energy, combined with the introduction of large-scale experiments in CO_2 pricing on its territory, constitutes a timely reminder.

It goes without saying that the impetus of such a dynamic comes at a price. The associated costs are primarily related to the effects of shifting investment toward innovative sectors in terms both of energy and of the environment. In Western service economies, progression and growth margins are shrinking, and the hope of making up lost ground with regard to traditional industrialization seems to be largely illusory. Nevertheless, the production of knowledge and mastery of green technologies is probably the only feasible way of ensuring sustainable growth in these economies. The implementation of this dynamic, however, tells us nothing about the social and ethical aspects of this growth. It is undoubtedly with respect to these dimensions that the political conditions for the implementation, or non-implementation, of the green economy will be determined.

9

Water, the Shepherd, and the Owner

A Choice of Green Growth Models

ANY CHANGE IN THE prevailing mode of growth entails costs, at least transient ones, and the primary question arising is usually "Who's willing to pay them?" The answer, though not surprising, may be somewhat discouraging: "Nobody!"[1] If this is really the case, then green growth is likely to remain a slogan without operational reach for the foreseeable future, and the ecological transition a pious wish whose fulfillment is endlessly deferred. Fortunately the question as it stands is poorly phrased because the idea of "cost" is ambiguous.

As we have shown, the inclusion of natural capital in the production function clearly results in a change in relative prices and a redistribution of costs. While involving no additional outlay, it makes explicit a use cost of natural capital that hitherto was to a greater or lesser extent deliberately hidden. By changing the relative prices of factors, the internationalization of this cost through the pricing of pollution results in a modification of the production function, which becomes more labor and capital intensive and more economical with regard to natural capital. If the economy is rigid, transition costs are likely to be high, leading to an accelerated displacement of physical capital and a rise in structural unemployment. If it is responsive and flexible, the new costs incurred will mainly be investment costs—precisely what the economic system is in dire need of in the short term to get it moving again.

By introducing natural capital into the production function, the greening of growth involves a rebalancing of production factors and a redistribution of income. These distributional mechanisms have been relatively little studied by environmental economists, whereas in practice they are one of the main difficulties of the transition to a green economy. It thus becomes essential to elucidate them clearly. Let us return to the shepherd and his sheep.

The Shepherd and Sheep Shearing:
Who Pays for the Water Used?

As we showed in chapter 8, if the stream dries up due to climate change, or if the water becomes unusable as a result of local pollution, there are likely to be consequences for the shepherd's productivity. The paradox is that pollution is a factor of production that produces its own limitations on the very conditions for the reproduction of production factors. It is this retroactive effect that we need to examine. Simply noting the decline in production is not enough for implementing a policy: we need to determine who will actually suffer the consequences and what should be done to improve the situation.

Applying the basic principle of the remuneration of factors of production, the shepherd must receive revenue proportional to his productivity, as do the water and the owner.[2] Indeed, the shepherd receives an hourly wage equivalent to the wool produced in an hour, amounting to two or five fleeces, depending on whether the wool is cleaned dry or with water. Because access to the water is free, the question of distribution does not arise: it is in the interest of both the owner and the shepherd to draw on the natural stock to increase their profit and wages respectively. Suppose now that the water resource becomes scarcer and has to be paid for at a price equivalent to half the shepherd's hourly wage. Who should pay? Conversely, investing in the quantity or quality of water may lead to an increase in overall production. Who then will benefit?

If the shift is entirely borne by the shepherd and his wage falls sharply as a result, then it is as though the owner of the sheep were also the owner of the water. This possibility would imply a green economy that could be termed capitalist, in which employees have to bear the burden of the transition. If the holders of productive capital are enterprising, such an economy might result in new capital accumulation highly conducive to growth. But it is very likely, in the absence of sufficient counterweight on the side of labor, that the growth process would soon be impeded by rising social costs.

If the shift is entirely borne by the owner, a symmetrical situation arises in which natural capital belongs to labor. While this option may be socially desirable, it is likely to quickly rebound against its beneficiaries, because the owner is highly unlikely to agree to the resulting fall in profit. One possible reaction on his part would be to divert his capital from the wool business and place it elsewhere. Another would be to fire the shepherd and automate the shearing and cleaning of the wool.

If we consider that natural capital is imbricated with the entire production system, the cost of using the water should impact both wages and profits. In this case, it makes sense to cut wages and profits in proportion to their previous contribution to production. If they want to recover their past earnings, the only short-term option for labor and capital owners would be to pass the bill on to consumers, a possibility that will mainly depend on the form of the corresponding goods and services markets.

This option is socially more acceptable. But is it conducive to growth? It all depends on the incentives to invest and the response of businesses. If the price of water is sufficiently high, capital owners can envisage obtaining the economy's average profit rate by investing in the means of production and therefore in water. The effect is very favorable to growth that is more respectful of the environment. Conversely, the risk is that the falling rate of profit diverts investment. Once the fleece market is no longer profitable as a result of water pricing, there will either be a decline in investment or a relocation of production.

This kind of issue may be found in a concrete example: the European Union CO_2 emissions trading system.

The Case of the European CO_2 Allowances Market

In deciding to create an emissions allowances market in its fight against climate change, Europe directly introduced CO_2 emissions into the production function. Allowances are not simply financial assets but function as a factor of production without which production cannot occur, in the same way as physical capital or labor. In order to produce, a firm subject to this system now requires capital, labor, and natural capital in the form of CO_2 allowances. Thus by introducing the value attached to the climate, there is a shift from a two-factor production function to a three-factor function.

Taking into account this third factor of production does not in principle affect wealth redistribution. During the first two periods of the carbon market, the vast majority of allowances were distributed to businesses free of charge. Doing so allowed the tricky question to be avoided as to which part of revenue—wages or profit—had to be cut back to fund the reproduction of natural capital. As is often the case, the failure to determine a tradeoff has been to the detriment of European public finances, which have foregone a steady source of revenue.

From the standpoint of (re)distribution, these choices have led to responses from the industries subject to allowances that in most cases have enabled them to increase the returns on their capital. In the case of the electricity sector, for example, Boris Solier estimates that between 2005 and 2010, European electric power producers were able to pass on around 60 percent of the cost of freely allocated allowances in the selling price,[3] thereby resulting in windfall profits, in other words, an increase in the rate of profit generated by the capture of environmental rent. In general, the introduction of carbon pricing through an allowances trading system with free allocation has not at all led to the distributional transfers that are often attributed to it by industry lobbies: the great majority of companies subject to quotas have in fact been net beneficiaries, and it is their customers who have been the main losers in relation to the transfers generated.

Has opting for free allowances been wise for growth? Channeling the value of carbon rent to companies amounts to a subsidy that could be worthwhile if they used it to invest heavily in a low-carbon economy. Imagine the amount of investment involved, both at the level of production and at the level of distribution of energy. Because this type of outcome has not occurred, most European governments have continued to subsidize the deployment of renewable energy and have decided to put up the majority of CO_2 allowances for sale by auction for the third period.

Once the public authority levies a carbon rent, which is perfectly legitimate given the nature of the atmosphere as a public good, the distributional impacts of carbon pricing need to be carefully considered. If the payment of CO_2 allowances by businesses occurs at an unchanged rate of profit, it will be as if the owners of productive capital own the environment.[4] The price of labor will be reduced in proportion, and we will find ourselves in a typically capitalist realignment. If firms pass on all of that value in their selling prices, they return de facto to the previous situation, but at the risk of losing market share or moving production outside of Europe, where emitting carbon remains free. Each of these possible forms of realignment delineates a different type of green growth.

The Illusion of "Window-Dressing" Green Growth

Let us now widen our investigation on the basis of the previous two illustrations. We first consider what happens if we persist in not introducing the value of the environment into the production function. In this case there remains a green growth program in name only, a crude illusion that allows society to slumber in the face of ecological risks.

If natural capital continues to be virtually free, which currently seems to be the norm in the real world, we have a form of "complacent" green capitalism, whether private or state based. Putting a green tax of a few cents on household products is not likely to change consumption or production behavior but allows a pseudo-awareness of the planet's environmental problems to be displayed. This is "window-dressing" or illusory green growth. Various minor taxes are introduced across the board in the name of the environment, while taking care that none of them is likely to overly disrupt production and consumption behavior. At the same time, appeals are made to people's civic sense to be better informed about environmental issues and to the goodwill of entrepreneurs and investors, who are supposed to be "socially responsible."

This type of compromise is easy to find in society: nobody wants the price of the goods and services they consume to go up. Similarly, few investors want the expected return on accumulated productive capital to be cut back because of the environmental impact. By not changing the relative shares of the various factors of production and not genuinely introducing the environment into the production system, such compromises will fail to give a fresh impetus to growth.

This illusory green growth reflects the de facto abdication on the part of politicians with regard to assigning a real value to the environment. As a way of concealing the true state of affairs from the public, there often follows a proliferation of voluntary initiatives by engaged citizens or companies spurred on by appeals to "corporate social responsibility." Such initiatives are often relevant and innovative but can in no way compensate for policy failures. To further lull public opinion, we find a proliferation of logos, labels, and other environmental certification displaying a "greenness" that purports to be everywhere but is in reality nowhere.

A good criterion for determining whether our society is or is not in the process of moving on from illusory green growth is to rigorously evaluate its progress with regard to effective pricing of the environment. From this standpoint, the economic and financial crisis resulted in a great leap backward, dramatically marked by the failure of the Copenhagen global climate change summit in December 2009, which was followed by the spiral of disintegration of the carbon pricing system in Europe.

"Capitalist" or "Socialist" Green Growth?

Suppose now that it is possible to price all of the components of natural capital, the reproduction of which is henceforth assured thanks to an environmental rent. The transition from a production function with two factors to one with three factors, associated with three revenues—wages, profit, and environmental rent—allows us to sketch the outlines of several types of growth that are distinguished from illusory green growth because the new system for pricing factors of production in fact changes the allocation of resources.

If a value is assigned to natural capital at the expense of labor, the outcome is green capitalism, whether market or state. The remuneration of natural capital is taken directly from labor productivity. It is then up to employees to try and increase their productivity if they are to avoid a fall in wages or an even greater distribution of wealth in favor of capital.

Another feature of this green capitalism is its promotion of the belief that productive capital alone is virtuous in terms of the environment (solar energy, wind energy, electric cars, etc.), while the labor factor, in its current form, finds itself in a kind of impasse and has to agree to reform and adapt to the new environmental order. The cost of adaptation is far from neutral and the mobilization of new productive capital—bearing in mind the end of the textile industry and the fragility of the steel industry—may well cost labor dearly. If new jobs are created, how many will have to disappear? Will the balance sheet be positive? Green capitalism does not change social relations but extends the life expectancy of existing social relations by imposing an environmental rent which is ultimately paid for by labor. The environmental costs of growth are transferred to social costs, which race out of control. The model is therefore likely to be called into question by those harmed by it. Consequently it is preferable to look for an alternative that can better balance environmental imperatives with those of social justice.

In contrast to green capitalism, one might aspire to green socialism, in which it is the return on productive capital that adjusts to natural capital. Such an alternative "socialist" growth model may be attractive for those who think that the realignment of productivity should be borne solely by physical capital. In this green socialism, growth involves the development of human capital but also the preservation, repair, and development of natural capital. To embark on this path, the return on productive capital has to be seriously and permanently adjusted in order to finance natural capital. Such an adjustment is not realistic in the context of the organization of existing global markets. It would therefore need to be accompanied by a new form of socialist planning that no longer directs resources toward the accumulation (or over-accumulation) of physical capital but instead directs them toward the accumulation of human and natural capital. The prospect of this happening seems unlikely, both for practical reasons and for want of a theoretical basis.

Imagine, however, that the conditions for the implementation of such planning were met. The work of the planner would soon become "mission impossible." To counter the impact of the collapse of investment that the falling rate of profit on physical capital would certainly produce, the shift of resources into natural capital would have to quickly raise productivity. This prospect seems somewhat unrealistic. To enhance the value of natural capital it is often necessary to start with substantial investments in physical and human capital. The production of renewable energy or efficiency gains in buildings or networks requires capital-intensive investments at the outset that will never be made if the initial return on physical capital is reduced too much. Furthermore, the significant productivity gains that can be generated by ecologically intensive agriculture also require significant investment in knowledge. It is for these reasons that this second path would probably soon lead to a slowdown in growth, or even to degrowth, through excessive reduction in the returns on productive capital.

In Search of New Equilibria

Midway between the two paths traced above, there is a third path where the remuneration of natural capital is effected by the simultaneous redeployment of revenues from labor and capital. If this redistribution occurs in proportion to the initial contributions to production of labor and capital, it does not change their relative share in total income. The application of such a rule would lead to advocating a transition to a green economy that is "neutral" in terms of income distribution. This less painful transition could appeal to politicians, given the perilous nature of change management. However, it is not based on any solid foundations.

- Socially, it presupposes that the initial share of the different revenues of factors is, if not optimal, at least satisfactory in contemporary societies—a supposition that it is increasingly difficult to sustain as the scourge of rising inequality gains ground.

- Environmentally, applying the rule of the "distributive neutrality" would involve sidelining a basic principle of environmental pricing, which aims to tax pollution in order to change behavior. The logic of the ecological transition must therefore take into account the initial contributions of the different factors to the deterioration of natural capital.

It is desirable, therefore, to use the lever of the green transition to seek new social compromises that will improve the existing distribution of revenues and make it compatible with the remuneration of natural capital. This view corresponds fairly well to Scandinavian approaches. Sweden and Denmark have, for example, introduced such advanced systems of environmental pricing within the framework of comprehensive tax reforms inspired by these general principles. In practice, the modalities of the redistribution of revenues between labor and capital are very hard to establish because they have to reconcile environmental goals with economic efficiency and social equity. One thing is certain: there will be no greening of growth without a savings effort weighing on the revenues of both labor and capital. No actor should be under the illusion that with green growth, shearing—to return to the shepherd metaphor—is free: any such expectations inevitably take the community back to the mirage of "window-dressing" green growth.

To introduce this new growth model centered on the protection of green capital, it is essential in practice to be able to measure and then price the damage done to nature by the economy. The next four chapters will provide an opportunity to deepen these evaluation questions lying at the heart of environmental economics. The first two concern the protection of biodiversity and the following two concern the action to be taken with regard to climate risk.

10

How Much Is Your Genome Worth?

FEW BOOKS HAVE HAD as much impact as biologist Rachel Carson's *Silent Spring*, published in 1962.[1] In it the author warned of the imminent disappearance of the bald eagle, the emblematic bird of American power that may be seen on the U.S. coat of arms during presidential public addresses. More generally, Carson inveighed against the devastating effects that the use of the pesticide DDT in farming and wetland management was having on wild bird species. The bald eagle was added to the list of endangered species in 1967 and the use of DDT was completely banned by federal law in 1972.

For most people, the threat to biodiversity is often seen through the lens of a few emblematic examples: the polar bear and the baby seal threatened by global warming and human wickedness; the panda, beloved of primary school children around the world, whose potential extinction they find extremely upsetting; the bear, whose reintroduction into mountainous areas in Europe reflects a collective concern for the wilderness; and the bald eagle, whose cause cemented the ecological movement in the United States during the

1960s following the publication of Rachel Carson's book. However, the question of determining how much the polar bear, the seal, the panda, or the bald eagle are worth makes little sense. Firstly, economists are relatively powerless in the face of symbolic or affective values. But more than anything, when a species is threatened with extinction, the loss of value in fact concerns an entire ecosystem that is impoverished or may be threatened with collapse.

For the economist, the real difficulty with biodiversity is that a value has to be given to the diversity of life that clearly does not lie in adding up the value of each species, but much more in their multiple interrelationships that provide us with so many services essential for life. The complex issue cannot be evaded if green capital is really to be incorporated into the functioning of the economy.

The Protection of Biodiversity, a Major though Complex Issue

The 1992 Rio Earth Summit gave rise to the signing of two major environmental agreements: one concerning the climate that led to the signing of the Kyoto Protocol in 1997, the other on biological diversity that came into force in December 1993. This latter agreement today numbers 193 signatories with one main exception, the United States, which has never ratified it. The agreement seeks to promote the conservation of the world's biological diversity and the fair sharing of the benefits of genetic resources.

Five years after the signing of the framework agreement on climate change, in 1997 the Kyoto Protocol was adopted and became the first international treaty comprising legally constraining objectives and economic instruments providing for the pricing of carbon. In terms of biodiversity, the path was much more circuitous. It was not until the signing of the Nagoya Protocol in 2010 that the agreement resulted in an implementing text, which largely adopted the general principles of the agreement and provided for the creation of the Intergovernmental Science-Policy Platform on Biodiversity and Ecosystem Services. This platform is expected to play an equivalent role to the Intergovernmental Panel on Climate Change (IPCCC), which since 1987 has been tasked with informing political decision makers of the state of scientific knowledge about climate change.

Since the Rio Earth Summit in 1992 and the International Year of Biodiversity in 2010, crowned by the Nagoya conference, summits have proliferated, reminding us of the dangers that economic development gives rise to for the diversity of ecosystems but without offering any forms of action to counter them so far. Thus Nagoya's conclusion is indisputable: biodiversity is declining, but we are failing to do anything to prevent it from happening.

The Pavan Sukhdev report, published during the Nagoya biodiversity summit of 2010,[2] plays a similar role for biodiversity as the Stern report does for the climate. Updating the earlier calculations by Robert Constanza,[3] Sukhdev estimates the services provided by nature to be worth $23,500 billion a year, or around 40 percent of global GDP. In other words, if a quarter of these services were threatened by mankind's damage to the diversity of ecosystems, 10 percent of global GDP would be at stake. We are dealing here with orders of magnitude comparable to those of the potential climate damage by the year 2050 estimated by the Stern report. The big difference is that we are unable to express this damage in a common unit, as is done for the climate.

To calculate the social cost of carbon, we simply divide all the damage caused by climate change by the volume of greenhouse gas emissions. The CO_2 equivalent functions as a universal yardstick allowing a unique price to be established because anthropogenic changes to the climate have a clearly defined cause: the accumulation of greenhouse gases.

Mankind's damage to biodiversity takes many forms. It stems variously from the artificialization of the environment induced by the expansion of human habitat, transportation infrastructures that slice up natural areas, discharges of chemicals and metals, agricultural and fishing practices, numerous recreational practices, and so on. What is more, climate change also contributes to the deterioration of biodiversity, even if up until now it is far from being the main cause. These multiple impacts on biodiversity make it impossible to develop a common yardstick equivalent to CO_2 for the climate. Although we can try and measure the losses to biodiversity by targeting specific areas or species,[4] we do not know how to take this richness into account in our economic choices because we desperately lack a simple, clear value on which to base decisions.[5] Overall, the Sukhdev report satisfactorily provides aggregate values that reveal the scale of the problem. But by what magnitude should this aggregate be divided to obtain a unit value, and hence a price or a cost, of use for decision making?

Is this a reason to give up trying to measure the value of biodiversity and the costs associated with its erosion? Such an attitude would be all the more regrettable because numerous examples show that the markets, sometimes in a rather disquieting way, take the measure of such and such an aspect of this diversity of life. Are you aware, for example, that your genome could soon have a market value?

What Price Do You Put on Your Genome?[6]

"The Human Genome Spared from Speculation" ["Le génome humain sauvé de la spéculation"], an issue of *Le Monde Diplomatique*, was thus headlined in 2002 following the announcement that all discoveries relating to the sequencing of the human genome were to be placed in the public domain. Does this mean that our genetic identity, our intra-species biodiversity, has a value in the eyes of certain people?

The human genome constitutes mankind's genetic identity, carried by the DNA present in every cell.[7] The sequence of the genetic code accounts for the fact that, even though everyone has the same DNA, we are all different. While this knowledge allows considerable progress to be made in the understanding and treatment of diseases such as cancer, it can also predict the predisposition to certain diseases and individual responses to treatment (such as its effectiveness and possible side effects) and allows drugs appropriate for each patient to be developed. The calling into question of third- and fourth-generation contraceptive pills underlines its importance. It is clear that if it is possible, through the identification of people's genomes, to anticipate reactions or pathologies, then this may potentially be of value in the context of insurance, job appointments, or food production.

If the insurance, food, and health sectors are interested in people's genetic identity, it is perhaps not for purely philanthropic reasons: the genome has a price, a market value.

To achieve, in 2003, the first complete sequencing of a human genome, $3 billion was required. Four years later, it cost $70 million to sequence the genome of Craig Venter (who was the first person to have performed the sequencing of the human genome) and then just $2 million for James Watson (the American geneticist and biochemist, and the co-discoverer of the structure of DNA). Today, the process has moved into the industrial stage and the race is on among several startups to sequence a genome for less than $100 by 2020. Although the United States assiduously keeps its hand on the fundamental knowledge, China has entered the fray, particularly with the city of Shenzhen, which has become a true global platform with hundreds of sequencers. Thus the same applies for the human genome as for computer or mobile phone microprocessors. The costs are plummeting and new markets can develop, especially in the areas of health and food, including the early detection of disease and allergies and of individuals' predispositions.

It may seem paradoxical to agree to pay in order to acquire intimate genetic knowledge of ourselves. But if you had the choice, now that the cost is low, are you absolutely sure you would not want to know your genetic code or that of your children? Are we not tempted to find out about our children's potential illnesses in anticipation of times when we might negotiate an insurance policy by showing their satisfactory genome? So how much are we willing to spend for knowledge of our genetic identity? €1,000? €300? €50?

If the human genome is the most fascinating, there are many more prospects for sequencing species of all kinds. Private and public companies are greatly interested in the DNA of plants for biofuels and bioplastics, and that of animals for their selective breeding. Such companies are consequently willing to pay for this knowledge so they can exploit it.

The genome, the constituent element of diversity among and within species, is not fixed but is constantly evolving and changing. The ending of this process when a species becomes extinct thus represents the loss of an opportunity, of a potential. And the more species there are to be studied, the greater the possibilities are for present and future development. This potential is thus one of the values that needs to be ascribed to biodiversity. Conniving in the loss of biodiversity has not only a moral cost but an economic cost too. It is society's responsibility, in the framework of a transition to a green economy, to incorporate these costs into its functioning. For this, various evaluation methods can be of help.

Even though such costs cannot be accurately measured, one thing is certain: they are not zero. Economists need to know how to measure the value of biodiversity so it can be taken into account in production, consumption, and investment decisions. In the absence of a yardstick of a global biodiversity market, two options may be envisaged. Either we attribute a purely instrumental value to biodiversity or we give it a value over and beyond its utility. In any case, we can only adopt a value based on what society accepts. Neither the polar bear, the panda, nor the elephant has intrinsic value—only the value attached to the disappearance of species and the loss of diversity of ecosystems. We can thus return to the measures proposed by anthropocentrism, the value mankind is willing to ascribe to nature and its diversity. To do this, we have no choice but to examine our willingness to pay for or to accept this good and to understand the bases of such consent.

Environmental Evaluation: The Usefulness of the Concept of Surplus

Current work on the value of biodiversity reveals a multitude of possible assessments, and research in molecular and cell biology is still in its infancy in terms of economic valorization.[8] The complexity of evaluation is all the greater because it is not enough to make an assessment species by species. Interactions between species (e.g., the impact of invasive species), environmental conditions (e.g., habitats, nutrients), and local differences (e.g., use of resources, acidification) also have to be taken into account. To assess the value of this stock, it would be necessary to give a value to all the sourcing, regulation, self-maintenance, and cultural services provided by the diversity of living species.

The same goes for biodiversity as for many other situations in economics: it is easier, and often more relevant, to evaluate a change in welfare than an absolute value of this welfare. Indeed it is simpler to measure what we lose or gain rather than the exact state in which we find ourselves. A deterioration or improvement in living standards is easier to measure than the standard of life itself. The construction of a highway near where we live makes us aware of our state of welfare before the highway actually comes into operation.

In economics, this shift refers to the concept of change in surplus. For a traditional economic good, a consumer or producer's change in surplus is obtained from a change in the price of the good. If the price of the good falls, the consumer's surplus increases (because with the same amount of money, he can buy more goods), while that of the producer decreases (for the same amount of the good sold, he receives less money). Symmetrically, a price increase reflects a deterioration in the consumer's welfare and improvement in the producer's. Economists use the term "Marshallian surplus" here.[9] At market equilibrium, this surplus amounts to the maximum benefit obtained by a consumer from buying a good. The idea is that through his buying behavior each individual consumer determines a form of demand function with a maximum acceptable price corresponding to his maximum willingness to pay. For the goods available in a market, this forms a demand function and, even if the surplus remains an individual notion, we can make it collective by observing the market price of the good in relation to its maximum price, at which point there is no further demand for it. The area between these two prices and the demand curve provides a measure of the consumer's surplus.

To calculate the surplus or its variation, the demand function for the good concerned needs to be known. If knowledge of the demand function is possible in the case in which there is an existing market, it is much more difficult in the absence of a market. And, as we have seen, for most environmental goods there is no market.

In the case of environmental goods, a further difficulty arises. In fact, the Marshallian surplus corresponds to a price change at constant income, that is, for a given budgetary constraint. For an environmental evaluation (air pollution, noise, loss of biodiversity, etc.), it is generally not the prices that vary—they do not exist—but the quality or quantity of the goods available (clean air, absence of noise, ecosystem services, etc.). Hence we need to refer to volumes rather than prices.

Karl-Göran Mäler, in an environmental context, rightly proposed considering a change in quantities rather than prices to define the concept of change in surplus.[10] Thus the following two questions need to be answered. Faced with a deterioration of the environment, how much are we willing to accept as compensation? And if the environment can be improved, how much are we willing to pay for it?

Another important point was introduced by John Hicks.[11] Rather than considering a change in welfare at constant income, he suggested that a change, especially in terms of the environment, should be seen as a change in income allowing the same level of welfare to be maintained. This shift of emphasis allows the preceding questions to be more precisely formulated. Faced with a deterioration of the environment, how much are we willing to accept as compensation to keep the same level of well-being? And if the environment can be improved, how much are we willing to pay for it and so remain at the same level of well-being? This approach introduces the idea that a change in the quantity or quality of the environment can be fully compensated by a change in our budget.[12] I agree to a highway, a wind farm, or a landfill in exchange for compensation. I agree to the transformation of a vacant lot into a natural park, but how much am I willing to pay for it? In both cases—willingness to accept or willingness to pay—my welfare must be the same as it was before the change. Symmetrically, we can take as a reference point the level of welfare after the change—after building the highway or creating the park. In this case, the following questions have to be answered. How much am I willing to pay for there not to be a motorway? And how much am I prepared to accept if the park does not exist? In both cases—willingness to pay or willingness to accept—my welfare must be the same as what it will be after the change.

Following Hicks and Mäler, we have two measures of surplus depending on the welfare reference point: the compensating surplus (with reference to the current level of welfare) and the equivalent surplus (with reference to the future level of welfare). The compensating surplus (referring to the initial situation) leads to measuring an individual's willingness to pay for the change in environmental quality produced. The equivalent surplus (referring to the final situation) is a measure of the consent to receive compensation to offset the gain in environmental welfare that the individual foregoes.

Intuitively, we understand that there will be considerable differences between what we are prepared to pay and what we are willing to accept, if only because there is a budget constraint in the one case and not in the other.[13] But more fundamentally, the choice of the measure implicitly involves a choice as to the ownership of the environment. Authors such as Robert Mitchell and Richard Carson argue that the best measure is the compensating surplus because the consequences of a change are generally assessed in relation to the current situation.[14] This approach amounts to implicitly assuming that agents have an ownership right over the initial situation and not over the final situation. Consequently, in the case of an improvement in their situation, people can only indicate a willingness to pay so as to benefit from it. On the other hand, in the case of deterioration, they must be compensated at the level of their willingness to accept.

Do we have, in principle, a right to clean air, clean water, and a healthy environment? If we believe we do, as is suggested by the U.S. Clean Air Act or the European Directive on water, then it is not the initial situation—polluted air and water—that should be the reference point, but the final (hopefully better) situation. It is thus the equivalent surplus that should be taken into account. Ultimately, the choice of reference point—the current situation or the future situation—in measures of changes of surplus and of their compensation comes down to determining the rights of all with regard to the environmental good. From this choice of reference point stem policies of redistribution or of the allocation of rights over the environment.

Some Examples of Evaluation

In the absence of a global biodiversity market that could assign a cost to the reduction in the diversity of ecosystems, economists are forced to use assessment methods derived from the theory of surplus. There are two broad types of evaluation methods for determining agents' willingness to pay to protect ecosystems or to receive compensation to accept their deterioration: revealed preference methods (or indirect methods), involving observation of agents' behavior in goods and services markets whose functioning may be affected by the state of the environment; and stated preference methods, allowing contingent evaluations to be made and based on direct methods for questioning agents.

The revealed preference approach refers to observations in existing markets. In the absence of real biodiversity markets, it is a matter of determining, by studying the behavior of other existing markets, whether observable trades and prices are partly influenced by environmental variables. House prices are typically related to the quality of the surrounding ecosystems, with many studies showing the impact of the quality of water, air, and green space.[15] The paper by Patrick Bayer et al.[16] reveals, for the average American household in a metropolitan area, a willingness to pay between \$149 and \$185 for a reduction of one unit of average ambient concentration of particulate matter (PM10). The analysis by Aurelia Bengochea Morancho shows that in Spain house prices in cities fall by an average of €1,800 for every 100 meter-distance from green space.[17] Methods of this kind measure "the market footprint of non-market goods," such as environmental goods.

Without going into the details of the different methods,[18] the basis is always the same, namely determining agents' willingness to pay for or receive in response to a change in the quality (or quantity) of the environment. The main revealed preference methods are the hedonic pricing method (particularly applied to property prices), the travel costs method (primarily for evaluation of natural parks or recreation areas),[19] the damage function method (in relation to health issues, and which estimates the statistical value of a life to be about 120 times GDP per capita),[20] and replacement methods (assessment of the cost of repairing the damage). It should be noted that replacement methods are the source of the idea of compensation markets, particularly that of compensation for loss of biodiversity.

The stated preference approach involves questioning people directly to find out about their willingness to pay or receive. If revealed preference methods are subject to many uncertainties and shortcomings,[21] stated preference methods have been extensively used. It is interesting to note that this approach, the main method of which is contingent valuation, was empirically validated in the *Exxon Valdez* affair.[22] Contingent valuation involves reconstructing a fictitious (contingent) situation or market and directly asking each individual about his or her willingness to pay for an environmental change. The only method enabling one to assess non-use values or the value of a project *before* its completion, stated preference techniques are increasingly used to evaluate projects that have positive or negative impacts on biodiversity.

On the basis of these methods, many studies, by considering hypothetical variations in natural resources, have tried to estimate the value of different environmental goods.

For example, Liam Carr and Robert Mendelsohn estimate a use value of between $700 million and $1.6 billion for the Great Barrier Reef in Australia.[23] The deterioration of the coral and the decline in biodiversity of this reef may result in the loss of more than $100 million per year, according to a 2009 study by M. E. Kragt, Peter C. Roebeling, and Arjan Ruijs.[24] In total, the report by Pavan Sukhdev puts the services rendered by the world's coral reefs (fishing, fish reproduction, tourism, etc.) at about $170 billion per year. In her study, J. K. Turpie offers an assessment of the effects of climate change on the willingness to pay for biodiversity in South Africa. In the absence of climate change, the willingness to pay is comparable to the national conservation budget, about $58 million per year. By introducing the irreversible adverse effects on biodiversity anticipated as a result of climate change, the willingness to pay rises to $263 million per year. The higher figure conveys an estimate of the full value of South

Africa's existing biodiversity. In France, the study by the General Commission for Sustainable Development assesses a willingness to pay about €15.20 per household per year to maintain biodiversity in forests.[25] With twenty-six million households in 2010, one obtains a total of more than €400 million that could theoretically be mobilized every year for a number of years.

These figures should, of course, be treated with caution. But as we will show in chapter 11, as a guide to action, it is better to have imperfect assessments than no assessments at all.

11

The Enhancement of Biodiversity

Managing Access, Pricing Usage

IF SEEKING TO GIVE a single price to biodiversity is misguided, it can and must be linked to a cost in the event of damage and to a gain for its restoration. Properly used, the assessment instruments developed by economists provide a better understanding of the incentives that must be put in place to curb the damage to biodiversity by combining the lever of access to fragile ecosystems with pricing of their uses.

Limiting access has historically been one of the prime instruments for safeguarding biodiversity, particularly through protected area policies. One of the first actions taken in this direction was the creation, by order of President Ulysses S. Grant, of Yellowstone National Park in 1872, the forefather of the world's protected areas. In its extreme form, access management can involve total restriction by forbidding entry into a protected area. Restricting or prohibiting access to certain ecosystems has an economic cost, namely foregoing the immediate use of a resource, which is fairly easy to measure. Its benefit is more diffuse and uncertain because of imperfect knowledge of ecosystem dynamics and displacement in time. For this reason, experience has led to extending simple restriction of access through mechanisms that remunerate activities favorable to the reproduction of the living species occupying the area. These activities are a new form of investment intended to restore the vitality of ecosystems.

Fiscal and parafiscal instruments (differentiated taxation of land according to its utilization, user charges), tradable allowances schemes, and payment for environmental services all allow a value to be placed on biodiversity by pricing some of its uses. Their applications, still in their infancy, can complement devices regulating or limiting access. Their deployment on a larger scale is likely to ramp up investments in ecosystem enhancement and make them a real driver of the ecological transition.

Protection of Fishery Resources: The Proper Use of Transferable Quotas

The case of fisheries resources has given rise to a substantial literature introducing biological considerations into economic reasoning since the 1950s. Originally static, the pioneering approaches of H. Scott Gordon (in 1954) and Milner Schaefer (in 1954 and 1957) show that the exploitation of free-access fishing areas leads to zero profit at equilibrium, whereas the resource is likely to generate a positive profit for lower levels of fishing effort.[1] In 1955, Anthony Scott introduced the intertemporal dimension of the resource, and in 1973, Colin Clark drew a parallel between the exploitation of a renewable resource and capital theory.[2] Decisions to exploit a renewable resource are ultimately very similar to investment decisions. The renewable resource thus has its own yield: the reproduction rate of the species is comparable to the return on investment. Open access leads to the productivity of the resource being neglected, excessive short-term exploitation, and, consequently, loss of future earnings.

On the basis of these analyses, the protection of marine biodiversity was originally conceived in the 1960s in terms of restrictive access to fishing areas. Such policies involved specifying an annual catch level. To this end, regulation was introduced with regard to fishing methods, such as net sizes and operating seasons.[3] These fisheries management methods were subsequently enhanced with a variety of instruments such as fishing quotas by species and area, which may or may not be transferable. If properly managed, transferable quotas can increase the effectiveness of action to protect an environmental good such as marine biodiversity. For this, it is necessary to set the total allowable catch at a sufficiently constraining level and to introduce a monitoring system that prevents incentives to fraud or the concentration of fishing rights.

The first transferable quota systems for managing fisheries were successfully implemented in New Zealand in the 1970s, followed by the Canadian sector in 1993 and the United States in 1996. The introduction of individual transferable quotas allows each individual operator to increase or decrease his quota by buying units from or selling units to other operators, with the total catch limit being determined by the regulator.

From a theoretical standpoint,[4] the transferable aspect of quotas amounts to introducing a tax per unit of catch, which accordingly increases the attention paid to each catch. One of the main problems of fisheries management in Europe stems from the nontransferability of quotas, thus further augmenting the political dimension of their allocation. In a system without transferability, and because of the shortage of fishing rights, operators no longer integrate the value for the protection of biodiversity.

The lever of transferable quotas could also be extended to other aspects of the fledgling economy of biodiversity protection. In most industrialized countries, land uses and urban infrastructure development are framed by complex regulations and local taxation. Introducing transferable allowances is one promising way to simplify the regulatory maze in terms of objectives for protecting biodiversity. For example, once overall objectives have been identified for a given territory with regard to areas to be reserved for the protection of living species and to areas for construction and the density of housing that may be built there, why not use the technique of transferable allowances to give an environmental value to different plots by organizing a market in which land usage rights may be traded?

Offset Markets

Pricing the uses of biodiversity may also take the form of contracts giving a value to investments that restore biodiversity and compensate for the original damage. The prototypes of such so-called offset markets first appeared in the 1970s in the United States. Their principle is to assess the damage to biodiversity caused by a particular economic activity, such as infrastructure construction, and to fund the restoration of at least equivalent biodiversity elsewhere.

In the United States, offset markets arose following the Endangered Species Act of 1973, which, in Articles 7 and 10, requires economic operators who cannot reduce their damage to the diversity of an ecosystem to restore an equivalent biological abundance in another location, under the supervision of the regulator. In practice, these offset operations have, since their inception, enabled nearly 300,000 hectares to be rehabilitated. They generate annual investment in biodiversity protection amounting to some $1.5 to $2.4 billion.

In France, the dedicated subsidiary of the Caisse des Dépôts introduced the first experiment in compensation for loss of biodiversity in 2009, under the 1976 law that requires agents to offset the harm caused by their activity. The system compensates losses to the diversity of ecosystems by means of infrastructure investments financed through the purchase of equivalent biodiversity units. The first site chosen was the abandoned area of Cossure, located in the Crau plain, which was originally characterized by a complex Mediterranean ecosystem. In 2011, the biodiversity unit purchased by companies or public authorities to offset their damage to biodiversity was estimated at €38,000 per hectare.

With growing awareness of issues of biodiversity, such instruments are also developing elsewhere in the world. The most serious estimates put the value of trade in this emerging market at between $1.8 and $2.9 billion in 2011 for investments that involved less than 100,000 hectares.[5] Faced with the global problem of the erosion of ecosystem diversity, it is but a drop in the ocean. Yet the mechanism is still only at the experimental stage, in that some damage is compensated and other damage is not.

The Internalization and Marketization of Services Provided by Biodiversity

Another experimental route to introduce the protection of biodiversity into the economy is to use the valuation methods described in chapter 10 to internalize some of its value through taxation and/or to marketize it through ecosystem service payments.

The bulk of taxation for maintaining the functioning of government and the welfare system is based on consumption, income, and labor costs. Local taxation is much more concerned with the use made of land, the impact of infrastructure built on it, and the funding of local public services such as waste disposal and remediation of water, whose connection with ecosystem diversity is obvious. Consequently, local taxation could become one of the preferred levers for the internalization of biodiversity.

In France, one of the first suggestions by the Committee for Environmental Taxation—set up in December 2013 to help the government "green" the tax system—concerned the fight against the growing artificiality and non-permeability of soils.[6]

Approximately 600 square kilometers are artificialized annually in France as a result of the extension of building and infrastructure, or the equivalent to a *département* every ten years, 90 percent of it at the expense of agricultural land. Slowing and then reversing this process is a priority that has been adopted across Europe so as to maintain the regenerative functions that living soil has on the natural environment. All the incentives that the Committee recommends implementing concern local taxation: adjustment of the planning tax, introduction of a levy for too low density, altering the property tax to discourage changes in the use of agricultural and forest land, etc.

This internalization of the value of biodiversity through local taxation is only very partial. It will be all the more incentivizing if its expansion is coupled with the development of payment mechanisms for ecosystem services. In the future, there will, for example, be much to gain by exploring this pathway for the management of the large water cycle for which the existing tax system is no longer really suitable.

Payment for ecosystem services involves valorizing the regulating services that biodiversity can provide to other actors or to the community. Estimates suggest, for example, that it would often be more productive to invest in forest management in the catchment area of rivers upstream than it would be to finance treatment plants downstream. But markets that would allow such investments do not exist. They cannot arise spontaneously, but instead require interventions from the local or national public authority to define the rules, as occurred when the City of New York decided to invest in the catchment area of the Hudson River upstream of the city and develop an ambitious afforestation program financed by economies made on water treatment downstream. European cities such as Munich and Evian have followed the same path.

One area in which the services provided by biodiversity are particularly important is that of agriculture (for a detailed analysis, see the INRA scientific expert report).[7] With 60 percent of its land area devoted to agriculture, France occupies a distinctive position in Europe and in the world. If historically agriculture and biodiversity have mutually enriched each other, changes in farming practices and landscape management—intensification and specialization of production; use of fertilizers, pesticides, and herbicides; land consolidation and simplification of landscapes; disappearance of non-productive areas; etc.—have led to a considerable loss of biodiversity.

Nonetheless, as the INRA report points out, there are numerous services provided to agriculture by biodiversity and various solutions are emerging. Its contributions are direct with regard to agricultural revenue, both through a potential gain in the quality of products and through better functioning of ecosystems by biological control (use of different species to combat unwanted species, pollinators, etc.) or by the provision of resources to plants (the fertility and physical stability of the soil, etc.). Alongside these direct contributions there is the role of biodiversity in maintaining water quality and in climate regulation, which make an indirect contribution to farmers' income and to overall welfare.

Various developments designed to make better use of biodiversity in agriculture are currently being tested: low-input cropping systems, simplified soil tillage, selective weed control, integrated production and organic farming, less intensive use of meadows, diversification, etc. While such practices may lead to lower yields, it seems that these losses can be offset by savings in terms of energy, fertilizers, and pesticides. For example, in orchards in southeastern France, pesticide reduction is being effected by planting orchard surrounds with grass, which constitutes a source of biodiversity favorable for pollination and integrated biological control.

With regard to the protection of biodiversity, the construction of ad hoc economic instruments, with a view to gradually introducing this value into the economy, is progressing steadily and in a decentralized way. This process is reflected in a multitude of local and regional values linked to investments to regenerate ecosystems, no two of which are ever identical and which differ in their quality and diversity in accordance with stresses of often very different origins. Unlike simple restriction of access, which constitutes a "passive" defense of biodiversity, these economic instruments are used to induce new investment that creates immediate economic wealth and employment. Although they are still very limited in scale, they form the seeds of a transition to a green economy. To gear up and cover those biodiversity values most distant from standard economic concerns, their deployment will also need to go beyond the narrow anthropocentric views so dear, because they are reassuring, to economists.

Biodiversity for Itself: A New Kind of Option Value[8]

The advantage, or disadvantage, of living systems is that it is very difficult to know how they will change and what their future potential for human use will be. As Bernard Chevassus-au-Louis has pointed out, with living beings it is always difficult or even impossible to predict what the outcome will be of the various decisions taken.[9] In his 2009 report he introduces the ideas of "remarkable" biodiversity and "ordinary" biodiversity.[10] As defined in the report, the former type of biodiversity concerns "entities (genes, species, habitats, landscapes) that society has identified as having intrinsic value and is based primarily on values other than economic," while the latter type has "no intrinsic value identified as such but, through the abundance of and the multiple interactions between its entities, contributes in varying degrees to the functioning of ecosystems and the production of services that are our societies find in it."

A narrow anthropocentric view will result in action being confined to the protection of "remarkable" biodiversity. In so doing, it will lead to a neglect of the extraordinary option value found in ordinary biodiversity, in which it is essential to invest even though its evaluation partly eludes economists' most sophisticated toolboxes. This option value is based on the fundamentally dynamic aspects of living systems, for most of which the focus is more on the preservation of the potential for change rather than on preservation itself.

Retaining such an option value in the protection of biodiversity leads to conceiving action in terms of biodiversity hotspots and ecological corridors. Some actions increasingly recommended by scientists, such as the reintroduction of the diversity of life into inner cities, may seem anecdotal. It may appear paradoxical that cities like New York or Paris are becoming a source of biodiversity. Yet the development of green spaces in cities on the one hand and urban sprawl and the intensive use of pesticides on the outskirts on the other hand justify making inner cities areas of biodiversity protection or even enrichment.[11]

The enhancement of these biodiversity hotspots is all the more important because they are potential sources of the spread of species. The creation of ecological corridors can also contribute to such propagation. It is thus important that infrastructure projects should include areas of biodiversity and external access. We need to make room for these sources of diversity even if we do not know ex ante their significance or future return.

The creation of artificial biodiversity areas is certainly open to debate. But it is undeniable that in modern society there is no place that has not been touched by man. No area in the United States or in France can be seriously seen as completely natural, and it is important not to confuse "biodiversity" and "natural." Through their behavior, farming, and uses of nature over the centuries, human beings have totally transformed their environment. These changes have given rise to new forms of biodiversity. Letting nature reclaim its "rights" at all costs in the countryside will, almost as surely as the construction of a highway, destroy biological hotspots. The creation by mankind of open areas such as meadows is an important source of biodiversity dynamics. By abandoning them to nature—and bearing in mind the phenomenon of entropy that is central to the principles of thermodynamics—we risk losing this diversity and tending toward a homogenization of species.

Of course with such a complex and shifting phenomenon as biodiversity, ranging from the diversity of individuals to macro-ecosystems,[12] it is not possible to measure future benefits, and it is very difficult to accurately evaluate current costs. We are thus obliged to accept the setting aside from the general application of classical economic regulatory instruments in the name of this new type of option value: the preservation of the potential for change constituted by the diversity of living species.

Due to the inherent complexity of living systems, evaluating mankind's adverse impact on biodiversity is a daunting task, and the introduction of economic incentives for its protection is still in its infancy. Advances in the pricing of this capital involve a series of case studies and experiments, the generalization of which is always difficult in the absence of a common standard similar to the one used by climate economists: the CO_2 equivalent, which the following two chapters will address.

Climate Change

The Challenges of Carbon Pricing

THE TERM "CLIMATE" COMES from a word in ancient Greek meaning "tilt of the heavens."[1] Climate designates average weather conditions (temperature, precipitation, wind, etc.) observed over long periods. These conditions may be local, regional, or global. At the global level, major changes have affected the climate system in accordance with geological eras extending over very long time frames, and in particular with the alternation of glacial and interglacial periods. Paleoclimatology reveals that such shifts were accompanied by significant changes in the amount of greenhouse gases in the atmosphere, especially CO_2.

The extremely rapid warming observed during the twentieth and twenty-first centuries is entirely new in kind: it reflects the first effects of the accumulation of human greenhouse gas emissions, which are distorting the exchange of energy flows between energy entering and leaving the atmosphere. Climate science is not able to predict with any certainty how the climate system will respond to this anthropogenic alteration of the composition of the atmosphere, but its diagnosis is that it is subjecting ecosystems and human societies to increasing risks.

The issue of climate change typically comes back to the problem of the free use of one particular environmental resource: the atmosphere, up until now used without limitation for the storage of all kinds of waste gases. Returning to Hardin's example of the feudal village discussed in chapter 5, we can replace the terms "village" and "communal pasture" with "planet" and "atmosphere." We then get one of the best possible descriptions of the problem of global warming: like the herdsmen taking their livestock to the pasture, industrialized societies have become accustomed since the beginning of the Industrial Revolution to using the atmosphere as a huge reservoir where they can freely release their greenhouse gas emissions. In doing so, they destroy its capacity to ensure the stability of the climate. Ending this free usage by means of carbon pricing is probably what will most contribute, in the coming decades, to the introduction of green capital into economic reality.

Climate Change or the "Tragedy of the Commons"

Climate economists have tried to attach a cost to this "tragedy" by estimating the loss of value resulting from the continuation of current greenhouse gas emission trajectories. The most comprehensive study, conducted by British economist Nicholas Stern in the context of his Review for the UK Exchequer in 2006, gives a figure in the range of a 5 percent to 20 percent loss of consumption by 2050.[2] This figure represents what it would cost tomorrow if we continue destroying our common resource today. Apart from this theoretical calculation, which is useful for alerting the public but does not specify what responses to make, can economists contribute to the introduction of appropriate actions to counter the risks associated with climate change?

Let us return for a moment to Hardin's villagers and examine how they can stop the destruction of the communal pasture. To save their common good, the villagers can take three types of action.

1) They can agree on alternative methods for managing the common good, for example by organizing rotations that limit access to the pasture. The American economist Elinor Ostrom has analyzed the various forms of local management able to increase the value of a common good, using the concept of "polycentrism," that we will return to when we discuss questions of governance. This approach, however, does not seem to be easily transferable to cases of planetary common goods, such as the climate.[3]

2) To facilitate the reorganization of the use of the communal pasture, the villagers can stop free access to it, for example by introducing a charge that its users will be required to pay. They will then be faced with two practical issues: How should the charge be set? And what should be done with the proceeds?

3) A third way to safeguard the fertility of the pasture is to distribute title deeds to the villagers and thereby create a land market. This was the route taken in England in the late Middle Ages, with the "enclosure movement" following the privatization of the commons. The term reminds us that one of the first decisions of the new owners was to enclose the fields so as to protect them against incursion by livestock, thus restoring their fertility and increasing agricultural productivity.

Environmentally, the first route corresponds to traditional regulatory approaches. These types of approaches are by far the most widely deployed in public policy, particularly in developed countries using a sophisticated array of regulatory instruments. Consider, for example, the 800 pages of the European REACH legislation regulating the use of more than 7,500 chemicals, the number of which is expected to triple by 2018. In the United

States, extensive regulatory power has been given to the powerful Environmental Protection Agency (EPA), which enacts various norms and standards. In France, the main measure resulting from the Grenelle de l'Environnement debates, intended to give new impetus to policies to protect the environment, has been the tightening of thermal standards in buildings. All these measures give rise to implementation costs (for the public authority) and compliance costs (for economic actors). In this respect, they clearly reduce free access to the common good. But they are not based explicitly on an environmental value that can be introduced into the economy either through taxation (regulation by price) or by the free market (regulation by quantity).

Regulation by Price: How Is the Level of a Carbon Tax to Be Set?

Because of the multiplicity of sources of greenhouse gas emissions, an international regulatory approach, such as was used successfully to combat the depletion of the ozone layer by CFCs and related compounds,[4] seems inappropriate for climate change. That is why the question of the use of economic instruments very quickly arose in the implementation of climate policies.

The first moves toward pricing greenhouse gas emissions were made in Europe in 1990, when the European Commission put forward a proposal to introduce a harmonized tax on CO_2, targeting the emissions of all major industrial sources. The rate of the tax was originally to be set at a modest level and then to grow over time. The expected proceeds would have contributed to the reorientation of EU policies toward innovation and support for research and development. This proposal was met with hostility by a number of Member States unwilling to give up their fiscal sovereignty and was finally abandoned in 1997 in favor of another instrument, namely tradable emissions allowances. The approach taken by the Commission nevertheless facilitated the initiatives of some European countries that have embarked on the carbon tax path: Sweden, Norway, Denmark, the United Kingdom, and, more recently, Ireland, Switzerland, and France. All of these countries were faced with two key questions pertaining to any environmental pricing: At what level should the tax be set? And what should be done with the proceeds? Provided it is used in a non-dogmatic way, economic theory here provides some useful guidelines.

The theoretical bases for environmental taxation were introduced by economist Arthur Cecil Pigou in 1920 in *The Economics of Welfare*. In this book, Pigou noted that one of the greatest market imperfections stems from the failure to take account of environmental externalities in the calculation of private sector costs and profits. He proposes correcting this imperfection through taxes that will be added to private sector costs in order to incorporate environmental values. He also suggests a method for determining their level: the tax rate should allow the environmental benefit that society will thereby obtain (the reduction of pollution) to be equalized with the cost associated with actions taken to reduce these environmental hazards. With an environmental tax, the agents causing pollution reduce their emissions as long as the cost incurred by their action is less than the tax. Once the cost of the action matches the amount of tax, they prefer to pay the tax rather than continue reducing their pollution. The optimal point is therefore reached when the marginal cost, that is, the cost associated with the reduction of the last unit of pollution, equals the environmental benefit expected by society. Economists usually refer to this benefit as the environmental dividend or the first dividend generated by Pigovian taxation.

The Pigovian principle is now used by all environmental research organizations under the name of the cost-benefit method. This type of approach is readily applicable to pollution that has identifiable local effects, such as pollution from urban traffic or the discharge of pollutants affecting the quality of a river. It is more difficult to apply in the case of climate change because of the complex interactions between the emission of a certain amount of greenhouse gases today and the cost of the damage the gas may eventually cause. The method is nevertheless used to calculate the social cost of carbon (SCC). The U.S. federal government puts this cost in 2015 in the range of $12 to $65, depending on the discount rate used—rather too wide to set a tax schedule![5]

In practice, a "cost-effectiveness method" is generally used, in which the level of the carbon tax is set on the basis of exogenously determined emission reduction objectives.

Without going into the details of the extensive literature on this subject, we can retain the simple idea that to reduce the risk of climate change, drastic changes need to be made regarding greenhouse gas emission trajectories. A rough figure, commonly cited from the work of the IPCC, is that global greenhouse gas emissions should be at least halved by 2050, having passed the emissions peak before 2020. Such trajectories would reduce the likelihood of having to contend with average warming of more than 2°C. With such trajectories determined ex ante, they can then be linked to a carbon price that minimizes the cost of achieving the objective. This theoretical price is known as the "shadow price of carbon." Stern (2007) puts the average worldwide shadow price at $25 to $30 per ton of CO_2 for the atmospheric concentrations of all greenhouse gas emissions to be limited to around 500 ppm. An extensive study carried out in France in 2009 under the direction of Alain Quinet uses a shadow price of around €32 per ton of CO_2 for 2010, increasing to €100 in 2030.[6] These two figures correspond to the objective of cutting greenhouse gas emissions in France by a factor of four. They are compatible with, though somewhat higher than, the shadow prices calculated using similar methods in other European countries.

This battery of shadow prices provides guidelines for legislators committed to the greening of taxation. Experience shows, however, that other considerations come into play and limit the ambition of carbon pricing. In France, following the recommendations of the Committee for Environmental Taxation established in 2013, the Finance Act introduced a carbon tax at a rate of €7 per ton of CO_2 in 2014, rising to €22 in 2016, or barely half of the shadow price of carbon. In 2010, Ireland adopted an introductory price of €15 per ton which subsequently increased to €20. In general, carbon taxes are set below their shadow price levels. The only exception in this respect is Sweden, which in 1991 was the first country to introduce such a tax, where the willingness to pay for climate protection has increased over time, as twenty years of experimentation have shown that increasing the carbon tax does not harm the economy—indeed quite the reverse.

Regulation by Quantity: Finding the Right Cap

Regulation by quantity involves pricing greenhouse gas emissions by means of a rationing mechanism that gives rise to a market value. Such a "cap-and-trade" system operates in two stages. First, the right to release greenhouse gas emissions into the atmosphere is constrained by establishing an overall emissions cap. Then once the cap is set, rights are distributed among emission sources subject to the rationing mechanism and a system for trading these rights is organized so as to give rise to a price.

Theoretically, it is immediately apparent that this carbon pricing system is exactly symmetrical to regulation by price. With regulation by price, the public authority sets the price, and the reactions of economic actors to this new cost lead to emissions being reduced by a certain amount. With a cap-and-trade system, the public authority sets the quantities and the allowances market reveals an equilibrium price, depending on the overall authorized quantity of emissions. Under perfect competition, that is, in an unreal world in which all actors have access to accurate information and no one can influence the equilibrium of the market, it is easy to show that the two systems are strictly equivalent. If we remove the supposition of perfect information, economic diagnosis becomes more complicated, as Martin Weitzman showed in a paper published in 1974.[7]

Advocates of emissions allowances systems generally rely on arguments that are more empirical than theoretical. They maintain that the cap-and-trade route faces fewer institutional barriers internationally and fits in well with a cost-effectiveness approach in which the emissions reduction target is set ex ante. The free allocation of allowances that generally prevails in the launch phases also allows lobbies hostile to carbon pricing to be mollified. Nevertheless, the first experiments in carbon pricing through tradable allowances led to the same conclusion as experiments in taxation: up until now they have resulted in obvious underpricing of climate risk.

At the international level, the Kyoto Protocol, signed in 1997, aimed to introduce a pricing system that would gradually extend worldwide. Its architecture was founded on the cornerstone of the international emission rights market among so-called Annex B countries[8] that jointly committed themselves to cut emissions of the six main greenhouse gases by a little over 5 percent between 1990 and the period from 2008 to 2012. At the end of the period, the conclusion as to the effectiveness of this device is clear: by imposing an emissions ceiling on Annex B countries in 1997, the agreement led to the establishment of a calculation and verification system for greenhouse gas emissions. This was certainly a step in the right direction. Yet as we shall see in more detail in chapter 13, it failed to create sufficient scarcity of emissions rights to make a carbon price emerge. The market for Kyoto units has remained embryonic, even rudimentary, and the system has not had an impact on global greenhouse gas emission trajectories.

The more mixed balance sheet of the European cap-and-trade scheme for CO_2 has not been wholly conclusive either. Originally created to help meet the European Union's Kyoto commitments, the operating rules of this system have been defined up until 2020, the target year of the climate and energy package, which sets the broad guidelines of European energy and climate policies. The system applies only to major industrial sources: power plants, steel plants, cement works, paper mills, glass factories, steam production, etc. In total, approximately 11,000 installations have been incorporated into the scheme, initially emitting two billion tons of CO_2. Before the start of the scheme, releasing these two billion tons was totally free for industry. The introduction of the carbon market restricted the unlimited right by giving a price to the right to emit a ton of CO_2. The total value of the cap consequently increased to €20 billion for a carbon price of €10 per ton and to forty billion tons for a price of €20. But the governance of the market revealed the public authority's profound incapacity to impose a cap that could give rise to a credible price over time. Prior to the start of the market, the main concern of the European and national authorities was that too strong a constraint would make the price soar at the expense of competitiveness. In early 2013, nine years after the market started, the situation was quite the reverse: the conundrum for the European Commission was to find a way of lifting the price, which had fallen to less than €5 per ton of CO_2. One obvious way would be to reduce the quantity of allowances in circulation or to introduce fewer. But in the framework of European governance, such a measure is very difficult to implement because the Commission cannot act without a mandate from the European parliament, and obtaining this can take several years.[9] As with the Kyoto market, the public authority has not succeeded in setting a sufficiently constraining cap.

Carbon pricing through quantitative regulation is also preferred in the United States and Asia. In 2010 President Obama just failed to pass the American Clean Energy and Security Act, a framework law on energy and climate that foresaw the development of a nationwide cap-and-trade scheme on a scale that would have quickly surpassed the European trading system. This setback was fatal to the voluntary carbon market developed in Chicago, but not to carbon pricing at the state level. The legally binding framework introduced by ten northeastern U.S. states, known as the Regional Greenhouse Gas Initiative (RGGI), got bogged down by an overly generous allocation and the incapacity of the local public authorities to remedy it effectively. The California market, launched in 2012, includes a floor-price mechanism designed to avoid such problems, but it has not been in place long enough for any conclusive lessons to be drawn. The same applies to the experiments launched in 2013 in five municipalities and two provinces in China, which are expected to lead to a national carbon pricing system in 2015. With South Korea launching its own carbon market in 2015, Asia is in fact set to be the world's principal laboratory for carbon pricing.

What Can Be Done About the Carbon Price? In Search of the "Second Dividend"

In practice, cap-and-trade emissions systems, like the attempts at carbon taxation, have proved disappointing, as they have not been implemented with enough conviction: governments do not set a constraint level leading to a carbon price in line with the desired emissions reduction trajectories because they are afraid that such a price would hamper the economy, adversely affecting its competitiveness and destroying jobs. They do not realize that wise use of the proceeds of carbon taxation would generate a "second dividend" of an economic kind.

Though Pigou provides the theoretical basis for the establishment of an optimal environmental tax, his approach sheds no light on the appropriate use of the value created by this tax, which forms a new rent—environmental rent—introduced into the economy. Similarly, while economists consider the most rational way of allocating rights in an allowances market to be selling them by auction, they are less forthcoming about the use that the public authority should make of the proceeds of these auctions. The public instinctively imagines that the "natural" purpose of a green tax is to finance policies for protecting the environment. In common parlance, an environmental tax is, moreover, much more a tax for funding the conservation of nature than a tax whose base is keyed to environmental pollution. Hence the many misunderstandings surrounding strategies for greening taxation: when economists think about changing the tax base, the public and its elected representatives argue that the proceeds from an ecotax should be used to fund policies to protect the environment. Work by economists on the "double dividend" may shed light on this debate. Their thinking applies equally well to green taxation and to the proceeds of the auctions used in paid-for emissions allowances allocation systems.

In Pigou's logic, the first dividend from the greening of taxation is the environmental benefit resulting from it. By construction, this alone justifies the introduction of the tax because the benefit that society derives in terms of welfare is greater than the cost it is willing to pay up to the point of marginal equilibrium, which equalizes the two terms. But depending on the use made of this tax, a second economic dividend may be obtained. For this to occur, the new environmental tax must replace another tax that has negative effects on business or welfare. Of course obtaining this double dividend is subject to numerous conditions.[10] We shall now review these.

In the view of Larry Goulder, who laid the theoretical basis for the analysis of the double dividend, to obtain the second dividend one has to presuppose an unchanged overall tax burden and a constant fiscal balance. Both of these assumptions are much more restrictive than they may seem at first. In particular, they imply that none of the revenue from green taxation is used to increase spending on environmental protection. In practice, such a choice comes up against the problem of a lack of understanding by the public and therefore of social acceptability. An ingenious way of circumventing this difficulty is to reserve part of the proceeds for tax cuts, which have an incentive effect favorable to environment protection. This technique has been practiced with some success by the Swedish legislature, which has offset some of the increase in environmental taxation by green tax cuts. A similar logic also applied in France with the introduction of taxation on purchases of high CO_2-emitting vehicles in 2008 by means of a "bonus-malus" system favoring clean vehicles, which was more acceptable to public opinion that a straightforward tax.[11]

Under the assumptions referred to above, the emergence of a second dividend depends on the respective economic effects of the environmental tax and the tax it is replacing. The reasoning, often presented in a complicated way, is actually quite simple: with a constant tax burden and a constant fiscal balance, the introduction of a green tax amounts to no more or no less than transferring a tax liability from a prior base to a new environmental base. This liability transfer brings an economic dividend if the distortion effect of the green tax is less than that of the tax it replaces. Conversely, if the green tax has a greater distortion effect, there is a risk of introducing a negative second (economic) dividend.

Three Possible Scenarios

Analysis of the double dividend shows the complexity in the practice of introducing environmental taxation. Assuming that the level of the tax has been correctly calibrated for reducing environmental damage and protecting natural capital, there are three possible situations in terms of economic effects.

1) In the first scenario, the green tax has negative effects on the economy that are higher than those of the taxes it replaces.

This is the case, for example, when the environmental tax adds to an already excessive tax burden or is abruptly applied to goods essential for production and there is no easily usable substitute. In this type of situation, society has to accept a net loss of economic welfare in order to increase environmental welfare. Consequently, the public is very likely to strongly oppose the introduction of a green tax.

2) A second scenario, termed the "weak double dividend" by Goulder, arises when the liability transfer induced by the green tax reduces, but does not eliminate, its economic cost. For example, if the green tax brings an environmental dividend valued at €100 and the loss of economic welfare is only €50 rather than €100, this means that the greater efficiency of the tax system has covered half of the environmental benefit. In a context of economic and financial crisis, it is not certain that the existence of a weak double dividend is enough, in the eyes of the public, to justify the introduction of an environmental tax.

3) The third scenario is one in which a "strong double dividend" is obtained. Such a twofold benefit arises when a tax that weighs heavily on economic welfare is replaced by a green tax without the same negative effects. In this case, a benefit in terms of economic welfare adds to the environmental benefit. In the real world, only this type of situation gives genuine political credibility to supporters of environmental taxation. However, the difficulties of establishing true carbon pricing through taxation reveal a considerable skepticism on the part of the public and policy makers as to the possibility of achieving this second dividend.

As noted by Mireille Chiroleu-Assouline,[12] a second dividend may also have a social component if accompanied by a gain in equity. In practice, this aspect is probably the most important because the introduction of a carbon tax leads to undesirable distributional effects insofar as it makes access to energy more expensive and disproportionally penalizes households with low living standards.

This question of fairness also arises with regard to the other way of introducing a carbon price, through a cap-and-trade system.

Lessons to Be Drawn from the Prototype Carbon Price

Such experiments in carbon pricing have constituted the first attempt to provide economic value on a large scale to part of natural capital. They show that the technical challenges related to the calculation and verification of emissions, at least for energy CO_2, can be overcome relatively easily.[13] The obstacles sometimes perceived regarding the transaction costs of this type of mechanism that would prevent any carbon pricing also vanish in the light of experience. Though this is an important lesson for the future, it is not necessarily transferable to other components of natural capital, particularly biodiversity.

Whether they involve a tax or a trading scheme, the experiments conducted so far point to the same conclusion: with few exceptions, the carbon price resulting from existing pricing systems falls far short of the price generally associated with climate risk. Such overcautiousness in terms of action may be accounted for by the fear of not controlling the distributional impacts in terms of competitiveness and fairness. The idea of an economic second dividend in addition to the ecological dividend is not in fact shared by the majority of policy makers, who view environmental rent as an additional burden on the economy. The extension of carbon pricing to elsewhere in the world on the basis of existing prototypes depends first of all on choices to be made with regard to income distribution.

Within each country, carbon pricing raises thorny issues of control of the distributional effects, on which stakeholders have difficulty agreeing. The fear that taxation weakens the competitiveness of the economy and destroys jobs is omnipresent, even though economic analysis suggests that if current levies on labor were to be transferred onto charges for pollution, this would have the opposite effect. In contrast, the example of Sweden shows that it is possible unilaterally to raise prices to levels far higher than the shadow prices calculated by economists when a social consensus is established around the type of distributional effects aimed for.

Internationally, climate negotiations have so far foundered on the inability to find a compromise on questions of redistribution between high- and low-income countries. From Kyoto to Copenhagen, regular attendees of major climate summits are well aware that the discussions are concerned relatively little with the climate and very much with how the costs and benefits of action on climate risk should be shared. Indeed, when it comes to attaching an economic value to the climate at the international level, a new question arises as to the distribution of environmental rent within the global economy.

13

International Climate Negotiations

ONE RECURRENT OBSTACLE TO the establishment of a carbon price at national or regional levels stems from the classic problem of "free riders."[1] The climate is a "planetary common good" whose deterioration risks causing major damage and whose protection requires immediate action. For each actor in isolation, there is no correlation between the level of effort he agrees to make in order to reduce emissions and the benefit he will derive in the form of less damage. Climate disruption is related to the global stock of greenhouse gas emissions, which is only weakly correlated with each country's annual emission flows. Furthermore, its impacts are distant in time, thereby encouraging each actor to pass on the full costs of climate change to future generations. In such a situation, it is in the interest of each actor to wait until his neighbors initiate action, the ideal position being that of the free rider who makes no effort while everyone else undertakes to protect the common good. Conversely, no player has an incentive to commit unilaterally for as long as he is not sure that others will follow as part of a wider coalition.[2] The paradox is that if no one does anything, everyone will suffer even greater consequences.

Faced with this free rider problem, Europe and the United States have responded in opposite ways up until now. High-minded Europe has always considered that the unilateral commitment of the high-income countries is likely to provoke an aspiration effect among other countries, which will spontaneously join a broad international coalition. In contrast, in 1997 the U.S. Senate adopted, by an overwhelming majority, a resolution opposing the ratification of any climate treaty binding the United States unless countries such as China and India were committed to similar efforts.[3] The resolution made it impossible for the United States to ratify the Kyoto Protocol and contributed to the stalling of climate negotiations. Yet the absence of effective coordination has led to alarming results: during the 2000s, global greenhouse gas emissions have accelerated and are further increasing our collective exposure to climate risk.[4] As long as the major players in the fight against climate change, such as China, the United States, India, and Russia, view their strategies as substitutable and not complementary, this trend will continue.[5]

The multilateral framework has nevertheless already proved its worth: the Montreal Protocol, signed in 1997, led the international community to virtually halt Freon gas emissions, the accumulation of which in the atmosphere was causing the destruction of the ozone layer. If an agreement was possible in the case of tropospheric ozone, why is it not also possible for greenhouse gas emissions? In this chapter, we examine the conditions needed for international climate negotiations to become a powerful catalyst for action. The role of economic instruments based on carbon pricing on an international scale is twofold: to encourage a growing number of actors to invest in the production and consumption of low-carbon goods and services; and to use this new environmental value to align the strategic interests of players attracted to individual strategies of climate risk circumvention.

The Framework for Climate Negotiations

The climate issue was introduced into international life in 1992 with the signing of the United Nations Framework Convention on Climate Change (UNFCCC). Two years earlier, the Intergovernmental Panel on Climate Change (IPCC) published its first assessment report, providing negotiators with reliable information on the state of scientific knowledge concerning climate change.[6] This linkage of the IPCC and the UNFCCC is an important aspect of climate negotiations. In the case of local pollution, exposed populations tend to spontaneously mobilize and put pressure on the municipality to do something about it. In the case of climate change—as with the destruction of the ozone layer—politicians have been alerted not by public opinion or environmental activists but by scientists, who have revealed complex causal chains between the atmospheric accumulation of greenhouse gases and climate disruption.

Coming into force in 1994, the UNFCCC was ratified by a large majority of countries, including the United States. It lays down three basic principles and a form of governance that provide the framework for international climate negotiations.

1) The first principle of the UNFCCC concerns recognition: in ratifying the treaty, each party acknowledges the existence of ongoing climate change and its anthropogenic origin. Legally, climate skepticism is thus prohibited for heads of states who have ratified the Convention! But for this principle to change decision making and help form coalitions, it is still necessary to ensure their adherence. This is the main function assigned to the IPCC, whose five Assessment Reports, published between 1990 and 2014, provide high-quality information for decision makers. In the United States, successive editions of the National Climate Assessment play a complementary role to the IPCC for domestic aspects.[7]

These advances in knowledge provided by the scientific community have not eradicated climate change skepticism. Science cannot convince militant climate skeptics, who deny the very existence of climate change. We are here in the realm of irrationalism and mystification, in a somewhat comparable way to the members of the Flat Earth Society,[8] who pursue the debate as to whether the planet is spherical or flat several centuries after the death of Copernicus. A more insidious form of climate change skepticism involves downplaying the risks of climate change because of its uncertainty in order to postpone any action until a later date. Uncertainty, however, is precisely central to scientific debate. This does not mean denouncing the absence of certainty regarding climate change in a sterile way, as often happens on TV shows, but involves raising new questions so as to advance our understanding of the complex interactions between human behavior and climate change. Uncertainties should not be denied but should be incorporated into the decision-making process while carefully weighing the various components of climate risk.

2) Secondly, the UNFCCC sets an ultimate goal, namely "to stabilize concentrations of greenhouse gases in the atmosphere at a level that would prevent dangerous anthropogenic interference with the climate system." The 1992 text is careful to specify what this level should be.

The objective of limiting average warming to 2°C compared to the preindustrial era was adopted in December 2009 at the Copenhagen summit and formally integrated into international climate agreements the following year at the Cancun conference. This objective in terms of temperature remains relatively unconstraining because it is not associated with a specific greenhouse gas emission and concentration trajectory. The work of the IPCC allows such trajectories to be traced, although based on a considerable number of assumptions and uncertainties. One simple idea to bear in mind is that unless emission trajectories alter, the planet will experience average warming of around 3°C to 5°C by the end of the century and such warming will continue during the following century. If the planet is to follow a less perilous route, global greenhouse gas emissions need to halve by 2050. Getting agreement by the major emitters on such trajectories is the central challenge of climate negotiations.

3) Lastly, the UNFCCC asserts the principle of "common but differentiated responsibility" in response to climate change. Here again, the principle of differentiation of the degree of responsibility depending on countries' level of development is incontrovertible. The UNFCCC provides a binary interpretation of this, which led the Kyoto Protocol, the Convention's main application text, to partition the world between high-income countries that are fully responsible for climate change and others that are exempted from participating in reducing emissions.

This bipolar view of the world, already questionable in 1992, is totally out of step with contemporary reality, in which emerging economies have shifted the center of gravity of the international economy by becoming the main drivers of the increase in emissions. Nor does it take account of the geopolitics of energy, with the importance of the bloc of oil producers and exporters, without whose participation no serious climate agreement can be achieved. These basic geopolitical facts are confirmed by the figures: of the top ten emitters of CO_2 from energy production, totaling about 80 percent of global emissions, five are emerging countries, four are industrialized countries, and one is an oil-exporting country (see table 13.1). An ambitious climate agreement must forge a new coalition among these major emitters by going beyond the binary interpretation of the principle of common but differentiated responsibility.

In addition, the UNFCCC introduced international governance, based on various technical bodies for the measurement, verification, and management of the instruments set up by the treaties. It also makes climate negotiations an ongoing process, requiring annual meetings of the supreme body of the treaty, the Conference

TABLE 13.1 Energy CO_2 emissions

	1990	2011	
	Gigatons	Gigatons	Cumulative global emissions (%)
China	2.43	8.67	27.8
United States	4.86	5.31	44.8
EU-27	4.13	3.59	56.3
India	0.59	1.81	62.1
Russia	2.34	1.74	67.6
Japan	1.07	1.19	71.4
Korea	0.24	0.61	73.4
Iran	0.19	0.52	75.1
Canada	0.42	0.47	76.6
Mexico	0.29	0.45	78.0
South Africa	0.29	0.45	79.5
Saudi Arabia	0.14	0.44	80.9
Brazil	0.21	0.42	82.2
Indonesia	0.15	0.41	83.5

Here we count the European Union as a single country because it negotiates with one voice at the United Nations. The table shows the top fourteen global energy CO_2 emitters. Taking into account CO_2 emissions associated with deforestation and other greenhouse gas emissions would put Brazil and Indonesia much higher in the ranking.

Source: International Energy Agency.

of Parties (COP), which is required to make its decisions by consensus of the 193 countries that have ratified the Convention.

Kyoto, Copenhagen, Paris, . . .

Three cities symbolize the key stages of international negotiation.

Kyoto (1997). The first stages of international negotiations were soon under way. Three years after the ratification of the UNFCCC, the Kyoto conference led to the signing of the Kyoto Protocol, the main application text of the Convention. The Protocol resulted in two important changes in international life: it committed, in a "legally binding" way, the industrialized countries to reduce emissions over the period from 2008 to 2012 by 5 percent compared to 1990; and it linked this commitment to a cap-and-trade system enabling these countries to trade emission rights, thereby giving rise to an international carbon price. A further feature of the system is the Clean Development Mechanism, which valorizes emissions reductions achieved by the countries of the South, otherwise exempted from commitments in the protocol architecture.

The promoters of the protocol originally intended extending it beyond 2012 by gradually incorporating other countries into the system. On realizing the impossibility of such a process, strongly advocated by the European Union, another type of political agreement was found in 2009 at the Copenhagen conference.

Copenhagen (2009). For the first time the emerging economies and the United States committed themselves to reducing emissions. But this major breakthrough on the extension of the geographical coverage was accompanied by a weakening of the monitoring system: under the Copenhagen architecture, the UNFCCC secretariat becomes a mere rubber stamp, recording the voluntary commitments submitted by countries without any homogenization of methods or a monitoring and verification system to ensure their implementation. Such an "à la carte" system has no chance of setting emission trajectories in line with the 2°C objective. In terms of economic instruments, the Copenhagen Accord is also a step backward compared to the architecture of the Kyoto Protocol. It simply records the financial transfer commitments from North to South—$30 billion from 2010 to 2012 and $100 billion per year by 2020—that have little real impact, as neither the origin nor the destination are specified, nor even the use of funds raised.

Paris (2015). In accordance with the customary procedure for climate negotiators, the Durban (2011) conference set the deadline of December 2015 to find an ambitious new climate agreement to bring into operation in 2020. In view of the number of topics up for discussion at the Paris conference, observers may be forgiven a degree of perplexity. The key issues, however, can conveniently be summarized on the basis of the three pillars of climate policy presented by William Nordhaus in his excellent book *The Climate Casino*.[9]

The first pillar is that preventive adaptation strategies strengthening the resilience of actors faced with climate change. These actions, implemented in a decentralized way, bring local benefits. International coordination here involves developing cooperative approaches to strengthen the adaptation capacity of the most vulnerable countries. There is no need for an international treaty to do this.

The second pillar consists of recourse, as a last resort, to so-called geo-engineering strategies consisting of artificially changing the climate regulation system in the event that the combination of adaptation and mitigation strategies were to fail—for example, seeding the sea with iron sulfate to increase its capacity to absorb atmospheric CO_2 or introducing aerosols into the upper atmosphere to reduce solar radiation. Such an approach raises many potential problems that it would be better to forestall by imposing very stringent international rules in terms of research, experiment, and, if need be, implementation.[10] For this, a treaty is required—but a new treaty lying outside the framework of the UNFCCC, because it will extend beyond its purview.

The central issue of the Paris conference concerns the third pillar, climate change mitigation, which involves acting with regard to greenhouse gas emissions. An ambitious accord in Paris would define a system that goes further than the one-legged Kyoto Protocol or the self-service system introduced in Copenhagen, in which everyone can nibble at what suits them. Such an agreement should pave the way for international carbon pricing that would put the world on a more reasonable emissions trajectory.

To initiate this change of dynamic, a set of incentives would need to be found to discourage the major emitters from continuing to act as free riders.

Possible Forms of an "Ideal" Agreement

Let us now try and outline the contours of the "ideal" agreement, in which a carbon price applies to every ton of greenhouse gas regardless of where in the world it is emitted. As noted in chapter 6, a uniform carbon price of $25 worldwide would generate an environmental rent of $1,250 billion at unchanged emission levels (50 billion tons of CO_2 equivalent, or 6.5 tons per capita in 2013). Impressive though it is, this figure would still only represent half the amount of oil rent. How might this sum be distributed in the global economy?

At the international level, the distributional effects of a single carbon price have been the real stumbling block in climate negotiations for twenty years. As Graciela Chichilnisky and Geoffrey Heal have emphasized, the differences in wealth between countries are such that the establishment of a uniform carbon price (whether in the form of a tax or an emissions trading scheme) seems impossible: a carbon price appropriate for the North will always be too high for the South, and one appropriate for the South will be too low for the North.[11] If we want to address this issue without proceeding along the very dangerous path of carbon prices differentiated by zones, it will be necessary to make massive resource transfers from North to South in order to guide the decisions of all economic actors under the right conditions.

On paper, the introduction of an international carbon price can be easily combined with such lump sum redistribution. Imagine that carbon pricing is introduced by means of a flat tax, the proceeds of which would be distributed in an egalitarian way to each country on the basis of the number of inhabitants. In distributive terms, such a mechanism is equivalent to a global cap-and-trade system based on an equal distribution of emission rights per capita. With unchanged emissions, such carbon pricing generates massive income transfers from industrialized to developing countries: a global flow of about $250 billion a year, twice the total public development aid. With a contribution of $115 billion, the United States would be the main loser, while India would be the main beneficiary, with an inflow of $135 billion. Such a system would be overwhelmingly approved by those developing countries most reluctant to join a climate agreement, such as India. But there is a strong likelihood that the high-income countries would be very resistant to taking that road.

In contrast to this first method, the so-called grandfathering formula could alternatively be adopted, in which historical emission rights are recognized and capped, and then reduced over time. A similar formula was used in the framework of the Kyoto Protocol, which has left the developing countries on the periphery of the agreement with the Clean Development Mechanism as a consolation prize. This architecture makes it very tricky to extend the original core of the coalition of high-income countries based on the acknowledgment of historical emission rights.

Various combinations of these two methods for allocating rights can be conceived that would redistribute winners and losers in the carbon pricing game. From an economic standpoint, this is the Gordian knot of climate negotiations that have been conducted continuously under the auspices of the United Nations since 1992.

To avoid these difficulties, it is tempting to envisage alternative systems. Thomas Courchene and John Allen have thus proposed introducing carbon pricing in the form of a tax imposed on the carbon content of goods and services consumed, along the lines of VAT.[12] Such a system deployed on the downstream stages of the economy is attractive because it neutralizes the risks of the undesirable competitive effects arising with progressive introduction and falls mainly on the consumption of high-income countries.

Concretely, its implementation would mean that carbon flows are tracked in the economy through microeconomic accounting, a distant prospect. The same applies to the proposal by De Perthuis (2010) to introduce carbon pricing through controlling emissions entirely upstream by capping the rights of fossil energy producers at the level of coal mines and oil and gas deposits. Such a system could incorporate fossil energy-producing countries into a coalition, but its implementation comes up against the problem of assessing the reserves available underground. In both cases, these routes are not likely to result in mechanisms that can be deployed on a large scale within the time frame required by the scheduling of international negotiations.

A Coalition of 193? Or of Three?

To move from an "ideal" agreement to an "effective" agreement that allows an international carbon price to be introduced, the number of players at the outset needs to be drastically reduced. From among the 193 countries involved in the UNFCCC, it is necessary to form a coalition of major emitters agreeing to combine their emissions reduction with a cap-and-trade mechanism that makes an international carbon price emerge. The two criteria for selecting the actors forming the initial coalition are size and their experience curve with regard to carbon pricing.

The size criterion shows that with just fourteen actors almost 85 percent of global emissions are covered, and that with the top three—China, the United States, and the European Union—as much as 56 percent are covered. A basic reality principle favors targeting the strength of the coalition rather than its extension to a larger number of players. Such a dilemma between numbers and strength is standard in the quest for environmental agreements.[13] In the case of a future climate accord, because of the concentration of emissions among a small number of countries, it would be most effective to build a solid core with a rather small number of players, which could later be expanded.

The experience curve criterion would imply selecting three players, all of them unilaterally committed to the carbon pricing route. The European Union was the first to introduce carbon pricing on a large scale, but it is struggling to find a second wind in its solo race. The United States introduced a cap-and-trade scheme to regulate federal SO_2 (sulfur dioxide) emissions in the 1990s and thus has the longest experience curve for this type of instrument. Because of the hostility of the Senate, such a scheme could not be developed at the national level for CO_2, but the government is in a position to learn from the experiences of individual U.S. states. Since 2012, China has been experimenting with regional carbon markets on the scale of municipalities and provinces, covering some 250 million people. As shown by the work of Wen Wang,[14] the design of these markets is often inspired by the European experience. Based on these experiments, the next stage of China's policy will be, beginning in 2015, to establish national regulation of greenhouse gas emissions together with a widening of carbon pricing.

The hard core of a future climate agreement could involve constructing a set of common emissions reduction goals from 2015 to 2020 by the three main emitters, with a long-term trajectory and a cap-and-trade system developed from the existing prototypes in these three groupings. Such a prototype of the international carbon market should, for practical reasons, initially cover only energy CO_2 emissions and construct a system of governance that takes on board all the lessons from the problems encountered to date by each of these main emitters for creating a carbon price. The very limited number of players may at first sight seem surprising. Yet it goes beyond a simple bilateral agreement between the United States and China, the route recommended in 2003 by Stewart and Wiener following the evident shortcomings of the Kyoto Protocol.[15] The success of this initial coalition will be measured by its ability to build a system open to other emitters, who should be encouraged to join it and thus consolidate it.

Countries not belonging to the initial coalition should of course continue to file their commitments with the UNFCCC, in compliance with the accounting and verification rules, even if these urgently need harmonizing and strengthening. These objectives are not initially shared in the emission rights trading system, though they will be in the following stages when the countries concerned successively join the original coalition on the basis of the twin criteria of size and their experience curve. If the size criterion is predominant, the most likely candidates are India, Russia, Japan, and Korea. But the experience curve criterion for carbon pricing should not be underestimated, because a country may not join the coalition unless it has already internally established the infrastructure needed for such pricing. From this standpoint, a country such as Korea, which is preparing to launch a carbon market system in 2015 covering its industrial emissions, is much more prepared than other large emitters such as India and Russia. How can such countries be encouraged to join the original three-member group? Generally speaking, the incentive will be all the stronger because the global climate agreement will simultaneously have mapped out positive pathways for the international coordination of action to address climate risks.

The Pathways of a "Positive Agenda"

Over the years, the agendas for climate conferences have tackled new issues, even though the negotiations may have been at a standstill or even backsliding in terms of coordinating actions to reduce greenhouse gas emissions. New topics, such as climate change adaptation, the transfer of low-carbon technology, and innovative financial mechanisms, have been introduced through ad hoc working groups without really opening up new perspectives.

The march toward increased cooperation in reducing emissions will be facilitated if these general categories are linked to specific questions that participants have raised, by suggesting they join concrete action programs to come up with solutions. By selecting one or two priorities in each area, there would be a gain in effectiveness and a shift from a strategic vision of substitution to one of complementarity. Let us provide a few examples to illustrate the point.

Regarding questions of adaptation to climate change, the IPCC Fifth Assessment Report upwardly revised its projections for the rise in sea levels and again emphasized the vulnerability of delta areas, the most populous of which are located in Asia and Africa. A program focused on defense strategies against this risk—with funding targeted to areas of highest vulnerability and the sharing of experience of the most innovative countries in finding solutions—would be more effective than general disquisitions on the respective merits of adaptation and mitigation.

Another topic emphasized in the IPCC Fifth Assessment Report is the vulnerability of food production in developing countries to anticipated climate disruption. Technically, an important way for farmers to adapt is to have living material (seeds and livestock) that can withstand changing climatic conditions. In its time, the "green revolution" contributed to major agricultural progress in these countries by initiating research programs around gene selection aimed at agriculture in the global South. Is it not the moment to launch a new program of this kind, under the auspices of the United Nations, to increase the response capacity of farmers in developing countries to climate disturbances?

With regard to technology transfer, much of the discussion introduced within the UNFCCC has focused on the question of patents and property rights, which risks hampering the transfer of low-carbon technology to the South. Though very important for the distribution of drugs for treating AIDS, this issue is not relevant for low-carbon technologies, the transfer of which can be very rapid, as shown by the relocation to China of companies producing solar panels. On the other hand, the deployment of carbon capture and storage (CCS) technologies faces many barriers. Yet the IPCC scenarios show that emission trajectories limiting the risk of global warming to 2°C require the large-scale deployment of these technologies. Rather than discussing patents, would it not be better to organize an international program of technological cooperation around CCS, drawing as much as possible on the lessons learned from pilot plants that could be installed in different parts of the world?

Are innovations likely to generate new funding? As the prospect of the rapid spread of carbon pricing recedes, proposals have multiplied for raising additional financing and finding ways to mobilize the $100 billion pledged at Copenhagen. We can certainly count on financiers to compete in devising unconventional products. Thus formulated, this question of funding makes little sense. If a climate agreement leads to a genuine prospect of pricing carbon, we will immediately see a variety of new forms of financing emerge for guiding the economy toward low carbon. Did the United States have any difficulty mobilizing the hundreds of billions of dollars needed to finance its non-conventional hydrocarbon revolution? Developing an Australian natural gas production and transportation chain destined for Japan involved an investment of around $50 billion. The major energy operators had no difficulty raising it. If their pricing system is changed by correctly setting the price of carbon, such investments will be redirected toward an energy transition compatible with the protection of the climate.

14

The "Energy Transition"

Not Enough or Too Much Oil and Gas?[1]

WHILE GOVERNMENTS STRUGGLE TO agree on ambitious policies to address climate change, a consensus seems to be emerging around the concept of "energy transition," now widely used in the world of energy. The advantage of the concept is perhaps its malleability, which in practice allows very different and even diametrically opposed policies to be justified.

In the United States, the energy transition consists primarily of reducing U.S. dependence on imported hydrocarbons from the Middle East, thus justifying the large-scale deployment of new drilling technologies to transform an oil and natural gas importer into a major unconventional hydrocarbon producer. In the most audacious (though unrealistic) scenarios, gas, the new domestic resource, is expected to replace conventional fuels in one in every two U.S. vehicles by 2020 and to extend the use of fossil fuels well beyond what could have been imagined only a few years ago.[2]

In Europe, the same concept justifies the development of poli-cies, ambitious on paper, to reduce greenhouse gas emissions, promote renewable energy, and encourage energy efficiency. Ostensibly, these policies are closely aligned within the framework of the "Climate and Energy Package" adopted in December 2008 by the Council of European Heads of State.[3] But once we dig a little, it appears that the label covers very different national strat-egies: in the name of the energy transition, Germany is exiting from nuclear power, the United Kingdom is seeking to return to it, Poland is seeking to acquire it, and France is organizing a major public debate. . . .

In the emerging countries, the energy transition is primarily aimed at ensuring adequate supplies to meet the needs of indus-trialization and the massive demand of households, of which an increasing fraction aspire to the living standards of the middle class in high-income countries, in terms of both housing and mobility. In the course of a just few decades, these countries are replicating a transformation—getting out of an energy system based on the majority use of biomass (primarily wood)—that took the high-income countries more than a century.

In fossil fuel-producing and -exporting countries, the energy transition is a concept that justifies the use of the rent pro-cured through the exploitation of underground resources for purposes of diversification of their economies, including devel-oping their own energy supply system. This latter dimension could boost economic growth in the Gulf countries, richly endowed with sunshine and windy exposed territories: Saudi Arabia thus plans to invest $100 billion in solar photovoltaic energy by 2020.

Given this profusion of possible interpretations, it is helpful to clarify the concept of energy transition. The most discriminatory criterion for doing so is the extent to which the price of carbon is taken into account in the energy system, as the example of the development of shale gas illustrates.

What Is an Energy Transition? The Lessons of History

Drawing on Vaclav Smil's key work *Energy Transitions*,[4] we can define an energy system as a complex consisting of three interrelated stages: a specific mix of forms of primary energy used, their transformation/storage/distribution chains, and their modes of end use, all of which are supported by hard and soft infrastructure with considerable inertia over time. The same triptych is found in Rifkin's more journalistic book on the "third industrial revolution," though without the historical depth or analytical rigor of Smil's book.[5] An energy transition thus involves not only a change in the energy mix, but also the evolution of this complex system from one dominant configuration to another. Smil's analyses reveal how we usually underestimate the length of these transitions and their structuring effect on the functioning of societies.

Smil identifies four energy transitions that have shaped the history of mankind. The first was the taming of fire, enabling the human species to acquire a major advantage over its competitors through the use of this energy drawn from biomass for cooking, heating, and later metal smelting. The second was begun by the Sumerians, who by means of irrigation first succeeded in increasing crop yields to feed domesticated animals and enable the population to lead a settled life. In terms of energy, this revolution added animal traction to human muscle strength. For plowing and the transport of crops, efficiency increased by a factor of four to six.

The third energy transition, beginning in England in the late eighteenth century, multiplied the amount of energy available thanks to the extensive use of an additional primary source, coal, which supplanted wood and the muscle power of men and domesticated animals around 1900 and established itself as the world's main energy source up until the mid-1960s. Often presented as the energy of the nineteenth century, coal nevertheless only played a significant role in the world energy system from 1880 onward and largely underpinned the industrialization of the twentieth century. The technical innovations permitting its use were available beginning in the mid-eighteenth century. Thus about 150 years separated such technical innovation and its large-scale deployment that would transform the economic system.

The fourth energy transition was based on a cluster of innovations that appeared simultaneously during the last two decades of the nineteenth century and allowed the harnessing of electricity (generation, transmission, and use in lighting and then industry) and the development of the gasoline- or diesel-powered internal combustion engine. Here we recognize two of the three major technical innovations identified by Gordon in his analysis of the growth process.[6] In fact, the spread of these innovations was the source of successive growth waves in the twentieth century. It led to falling prices, making possible the extensive use of new goods and services such as the electric light bulb. Based on the observations of Roger Fouquet, the falling cost of lighting caused by the shift from candles to kerosene lamps and town gas, and then to the incandescent bulb, rivaled the present-day fall in the price of computer memory.[7] Numerous goods would follow, from the washing machine—the first models became available in 1907 in the United States—to the computer and the various types of equipment used for transport. Their accumulation transformed people's ways of life and created the conditions for the mass consumption that led to the rapid growth of the last fifty years. Here, too, many decades separated the technological innovations, mostly introduced before 1900, and their impact on growth, fully manifested only after 1950.

In terms of energy, the fourth transition resulted in two major changes. First, it increased energy consumption to an unimaginable extent through the mobilization of three primary sources—oil, coal, and gas—providing 80 percent of the world's energy. In 2013, consumption was about two tons of oil equivalent per capita (7.5 in the United States, 3.5 in Europe, 1.8 in China, and less than 1 in India and sub-Saharan Africa). At the beginning of the last century, the figure is estimated to have been 400 pounds of oil equivalent at most. Second, this system is organized around a sophisticated network of hard infrastructure related to the extraction, processing, and distribution of this energy. In addition, there is soft infrastructure comprising the markets and institutions required for the regulation of what has become a highly complex system.

The first four energy transitions were the result of human ingenuity, enabling the limits imposed by the scarcity of usable energy to be overcome. As we have seen, such ingenuity led to remarkable efficiency gains, reflected in the ever-increasing use of energy since the beginning of the Industrial Revolution. As noted by Jevons in his celebrated essay on coal, "It is a confusion of ideas to suppose that the economical use of fuel is equivalent to diminished consumption. The very contrary is the truth."[8] The lowering of its cost was in fact produced by its increased use.

The fifth energy transition will free the energy system from its addiction to the three fossil fuels dominant today. As Smil (2010) points out, there are two reasons to pursue it: "Concerns about the long-term effects of global climate change, and worries about the rapidly approaching depletion of low-price, high-quality fossil fuels."[9] If we emphasize the second and focus on the risks of supply disruptions or price increases of fossil fuels because of their scarcity, we remain within a traditional Hotelling-type approach. If we take climate risk into account, we need to go beyond this approach and introduce a new value into energy markets: the value attached to climate protection, that is, the price of CO_2.

The Hotelling View of the Fifth Transition: Beyond "Peak Oil"

If it is a question of finding the optimal way to respond to the growing scarcity of the stock of fossil fuels remaining to be used, the toolbox developed by Hotelling is entirely appropriate. In this case, the way markers of the route are provided by energy prices will express these relative scarcities. In the very long term, the increasing scarcity of fossil fuels will result in an increase in their prices due to the increased cost of extraction and processing and of the forming of a rent associated with the gradual depletion of the stock. This rise in prices firstly encourages energy efficiency and progressively makes profitable investments in renewables. But it also boosts greater investment in exploration, and the energy transition becomes compatible with an increase in the use of fossil fuels once new deposits are discovered. Here the very long-term outlook for rising scarcity rent is blurred by the transient abundance resulting from new discoveries. Such was the case with the oil price countershock of the 1990s, with the ramp-up in development of new non-OPEC oilfields, and is currently the case with shale gas and unconventional hydrocarbons more generally, which are transforming the global energy landscape.

The choices made by North America or the oil-exporting countries come within this first approach. Here it is a matter of replacing or supplementing the fossil fuels currently used with other primary sources as these become less expensive. Thus it involves continuing to think within the framework of a system bounded by a set of scarce resources, in which the amount that can be extracted in the future is currently unknown given the evolution of economic and technical conditions. In accordance with the Hotelling view, the scarcity of stock generates a rent transmitted by energy prices to producers, who have a powerful incentive to increase their investment in exploration. The experience over the last fifty years suggests that we are far from having exhausted all known and unknown seams, not to mention the potential represented by methane hydrates, which probably contain more energy than coal, oil, and natural gas together!

Pursuing an energy transition based on this approach will eventually result in a reduction of greenhouse gas emissions. The path will be determined by the complex effects of higher long-term fossil fuel prices on supply and demand, with transient phases of fossil fuel abundance reflected by downward price cycles.

Given the high inertia of the energy system, as pointed out by Smil, and the powerful incentive to use oil and gas revenues to increase exploration investments, there is no chance that this type of trajectory is consistent with what is required to reduce climate change risks. The carbon stock locked up in the form of fossil energy underground is about four times the amount in the atmosphere, mostly as carbon dioxide. The extraction of merely a quarter of this underground stock would double the atmospheric concentration of CO_2, with major consequences for the climate system.[10]

The Energy Transition and Carbon Pricing

If the aim of the energy transition is to find an optimal path in terms of greenhouse gas emission trajectories for protecting the stability of the climate, while taking into account the constraints of competitiveness and energy security, it is necessary to abandon the traditional approach in favor of a shift that is part of a broader ecological transition. Here the strategy to adopt is quite different because there is far too much carbon in the stock of fossil fuels in the ground in terms of what the atmosphere can absorb without risk to the stability of the climate. The Hotelling logic prevailing in the energy markets creates powerful incentives, through oil rent, to recover an ever greater proportion of the fossil energy stock.[11] It is therefore necessary to "force" the transition by introducing a new value into the equation: the scarcity of the atmosphere expressed by the price of CO_2. It is important to introduce this price at the start of the transition because global warming is linked to the stock of greenhouse gases in the atmosphere, not to the annual emission flow. The cost to the climate of a ton of CO_2 released today is higher than a ton emitted in fifty or a hundred years' time.

Contrary to what is sometimes claimed, the price of CO_2 is not at all equivalent to a simple rise in fossil fuel prices. A carbon price allows the right incentives to be transmitted, both on the demand side (energy efficiency and the replacement of fossil fuels by renewables) and the supply side (reorientation of capital flows toward low carbon). An increasing CO_2 price trajectory over time thus eventually produces a decrease in the price of fossil fuels, for which there is less and less demand in the market. Above a certain level (currently estimated at sixty to ninety euros per ton of CO_2), the carbon price also encourages the use of new techniques for carbon capture and geological storage[12] that could enable fossil fuels to be used in the future with virtually no greenhouse gas emissions.

The essential feature of an energy transition based on correct CO_2 pricing is that it will lead either to the ending of the use of fossil fuels that have become too expensive or to burning them using only technologies that reduce their energy efficiency and avoid the associated release of carbon dioxide into the atmosphere. Much less carbon is therefore transferred to the atmosphere from the reservoir formed by coal mines and oil and gas deposits. Otherwise, without generalized carbon pricing introduced either through allowances markets or taxation, any energy transition strategy results in far too much carbon being released into the atmosphere.

We can usefully illustrate this point with a specific case: the revolution constituted by the development of non-conventional hydrocarbons in the United States, especially shale gas, that have profoundly changed the global energy landscape.

Shale Gas: Coal Substitute or Supplement?

The gas used for energy purposes, most of it methane, has three main origins. The most important is extraction from conventional deposits, and the product is referred to as natural gas. The same gas may come from the fermentation of organic matter, for example by being recovered from landfills, compost farms, or methanation units. It is then known as biogas. Finally, it may be recovered from gas pockets present in rocks using new extraction methods—horizontal drilling and hydraulic fracturing. It is then called "parent-rock hydrocarbon," commonly known as "shale gas." These three gases differ little in terms of their energy properties but greatly in the way they are produced.

Upon combustion, each of these three gases produces an average of 50 percent less CO_2 than coal. Their climate impact, however, can really only be assessed by following the emissions throughout the production chain. As well as releases into the atmosphere from the combustion of natural gas, there are also releases resulting from methane leaks during its extraction[13] and transport ("fugitive emissions") and from gasification-liquefaction processes when the gas is liquefied for transport by sea.[14] The recovery of biogas, on the other hand, lies within the short carbon cycle and allows fossil sources to be economized. It is positive with regard to greenhouse gas emissions because the carbon released during combustion has a much lower greenhouse effect than the methane that natural decay would have emitted. Lastly, the shale gas used in the United States avoids transport by LNG tanker, but the amount of fugitive emissions depends very much on the circumstances of its extraction. When the environmental conditions are very lax, as was the case in the early extraction fields, these fugitive emissions may exceed the gains resulting from the replacement of coal by shale gas in power plants.[15]

If we express climate in terms of the price of carbon, we obtain a clear order of preference. Because the biogas cycle is positive for the climate, it becomes the most economically advantageous source as the carbon price rises. To decide between natural gas and shale gas, we need to consider their respective fugitive emissions, which depend on extraction techniques, the state of pipeline maintenance, and liquefaction.

If the value of the climate is not included in prices and costs, the order of preference depends on conventional production costs, giving the United States a massive advantage with shale gas, whose development is currently transforming the global energy landscape. Without a carbon price and with pricing almost nonexistent for other environmental impacts associated with its extraction—mainly the natural areas and spaces used, the underground impact of fracturing, and seismic risks—the production costs of U.S. shale gas are usually estimated in the range of $5 to $7 per thermal unit.[16] Its accelerated development has reduced the equilibrium price in the gas market to below $5 per thermal unit in the United States, compared to $9 to $10 in Europe and up to more than $15 in Asia.

Due to the abundance of U.S. reserves, gas development is the main vector of the country's energy transition. Gas has started replacing coal in power plants. It is the source of a major relocation of the fertilizer and chemicals industry, using gas as feedstock. Its applications in transport are rapidly being developed with the stated aim of reducing the share of imported oil. This economically coherent strategy is also presented by its promoters as virtuous from the standpoint of climate due to the decline in the use of coal. The 10 percent fall recorded in CO_2 emissions from energy between 2005 and 2013 in the United States was, they say, mainly because of the substitution of gas for coal in the electricity-generating industry.

This account is misleading in that it is only part of the story. Coal from Wyoming and the Appalachians that is no longer used by U.S. power stations is by no means an automatic plus for the climate: some of the surplus made available has already been shipped to Europe and Asia, where it results in a symmetrical replacement of gas by coal because the relative prices of these two energy sources are more or less inverse.[17] Adjustments in the energy markets in their traditional operational mode spontaneously result in this type of pattern. Thus it is not shale gas *or* coal, but shale gas *and* coal, that this "made in the USA" energy transition presages.

In the absence of any real pricing of greenhouse gas emissions internationally, shale gas is added to coal resources used rather than replacing them, as the latest scenarios from the IEA precisely indicate. The inclusion of unconventional gas deposits has led to a massive upward revision of world reserves, from 60 to 230 years at current extraction rates, according to the IEA—equivalent to quadrupling the estimated amount of greenhouse gases that could be released into the atmosphere as a result. Without the marker of a CO_2 price, it is almost certain that most of these new sources will be added to coal, not substituted for it, as has already begun happening with U.S. coal exports. This type of energy transition leads straightaway to scenarios in which far too much fossil energy is used, given the climate risk. Herein lies the main lesson from the development of shale gas: in the absence of pricing of environmental and climate externalities, energy markets do not allow the appropriate decisions to be made for the current system to transit to a low-carbon target.

In view of this situation, how is Europe reacting? In a word, very unevenly. For obvious strategic reasons, Poland and Ukraine were the first countries to grant exploration licenses to major oil companies, to the annoyance of their Russian neighbor. A majority of other European countries are preparing to allow drilling so as to get a better idea of their reserves. Five countries have so far refused, including France, where exploration licenses originally granted by the government were subsequently withdrawn. The French decisions were based on local environmental considerations and will probably be revised once new drilling technologies using less water allow the country's shale formations to be explored at lower risk of local groundwater pollution.

Instituting an energy transition strategy that takes serious account of the risks of climate change entails pricing carbon accordingly and making unpopular decisions in terms of energy prices. When one is running for office, it is much easier to tell voters that taxes on gasoline or gas will be reduced so they can maintain their purchasing power. In actual fact, candidates should be promising voters precisely the opposite. For as we shall see, the issue of energy pricing is an inescapable aspect of the type of transition aimed for.

15

The Inescapable Question
of the Price of Energy

HISTORICALLY, ENERGY TRANSITIONS HAVE been triggered by falls in the prices of the energy sources used by society, resulting from technical innovations applying both to extraction processes and to distribution and end uses. The cumulative price declines, from upstream (primary energy prices) to downstream (price of usage or services provided by energy), have caused huge ratchet effects, greatly increasing the amount of energy used and hence boosting economic growth. The third energy transition, for example, beginning in the late eighteenth century, was triggered by an original technical innovation, the capacity of the steam engine to convert heat into mechanical energy. It made the early stages of the Industrial Revolution possible thanks to a cumulative decline in energy prices resulting from efficiency gains, spreading from coal mines to railways and the textile industry. The increase in the prices of the initially dominant energy sources—wood and muscle power—played virtually no role in precipitating this transition.[1]

Conversely, the fifth transition to a low-carbon energy system entails that the value of the climate be integrated at all levels of decision making through the generalization of a carbon price. The introduction of a carbon price has the effect of raising the price of fossil fuels: a barrel of oil bought at $100, for example, would cost $108 if the price of a ton of CO_2 released into the atmosphere was $20.[2] If the price of a ton of CO_2 rises to $200, a barrel of oil would cost $180.[3] And because fossil fuels account for 80 percent of energy sources, this automatically leads to higher energy prices overall. Though there may well be a consensus on the theoretical utility of a carbon price, it is more difficult to take the political decision to introduce it because it results in access to energy becoming more expensive.

To make the energy transition a success, prices must not only express the relative scarcity of different types of energy in conventional markets, but also the cost of their environmental and health hazards. For this shock to be beneficial to the economy, the rise in cost has to be absorbed by a pair of mechanisms: accelerating efficiency gains, thereby enabling the amount of primary energy consumed to be reduced, and a cumulative reduction in the cost of non-emitting sources. The winning economic equation of the transition is the one that reduces the bill for correctly priced energy through increased efficiency and the new competitiveness of non-fossil sources.

Energy Prices and Energy Costs: The Key Role of Efficiency

In the energy system, the introduction of a CO_2 price generates various incentives that encourage the reorientation of investment and consumption toward low-carbon systems. Consumers see the cost of using fossil fuels rise and will therefore economize and try to substitute other, non-emitting energy sources. At the same time producers are prompted to invest in the search for new, non-fossil energy sources or technologies for capturing CO_2 from the combustion of fossil fuels to prevent it escaping into the atmosphere, and industry as a whole will seek to make improvements in energy efficiency, a factor that plays an often underestimated role in energy transitions. To bring these incentives to bear in an optimal way, carbon pricing should be introduced gradually, following trajectories that need to balance various constraints.

Because the initial higher cost of fossil fuels applies to 80 percent of the primary sources used, its impact, all else equal, is to increase the total energy costs for the end user, even if there is an attempt to subsidize the remaining 20 percent (biomass, nuclear, hydro, and other renewables). The introduction of a price at a high level, especially if it is not implemented on a global scale, therefore runs the risk of creating asymmetry effects and causing carbon pricing to be rejected by society. Moreover, for this new "carbon price" signal to impact investment decisions effectively, it must be calibrated to increase over time as the least costly emission reductions are made. The economy must thus be prepared to face continuing price rises and not simply shocks limited to the short term. For the energy transition to remain a factor conducive to growth, it is consequently necessary to find powerful mechanisms that counteract the recessionary impact of rising fossil fuel prices. These are of two kinds: the rapid development of cheaper non-carbon energy sources as an alternative, and the acceleration of productivity growth in energy services increasing efficiency of use.

Scenarios attempting to trace the contours of the fifth energy transition show that while both types of adjustment should be implemented in parallel, the acceleration of energy efficiency gains plays a major role in the early stages of transition, when the learning effects have not yet been sufficient to disseminate the deployment of lower cost, non-fossil sources on a large scale.

Will the deployment of cheap non-carbon sources be able to quickly absorb the shock of rising fossil fuel prices? The three main existing sources are biomass, nuclear energy, and the various renewables, such as hydro, wind, solar, etc. The development of biomass energy has great potential, but its large-scale deployment comes up against the well known constraints of the risk of depleting the natural environment, competition with food needs, and logistical bottlenecks. Technical innovations will enable new sources to come into play—woody material, algae, etc.—but we are still far from the stage of large-scale deployment. The civil nuclear industry has, since its launch, faced increasing costs due to the proliferation of safety issues and the race to make new plants ever more gigantic. Hydropower offers limited opportunities in developed countries, where most favorable sites have already been put into operation, but greater potential in developing countries. The biggest growth area lies in flow energies, predominantly wind power and solar thermal energy. Thanks to technical innovations, their production costs are tending to decline. Under the best conditions, the costs are lower than those of competing fossil fuels, as is already the case for onshore wind and solar thermal. But these sources are characterized by a lower energy density compared to fossil fuels,[4] very irregular geographical distribution, and a mismatch with existing energy networks. It will therefore take time for these cost reductions at the level of production to have a significant impact on consumers by reducing their energy bills.

It is for this reason that achieving an energy mix compatible with the goals of climate stabilization depends on an acceleration in efficiency gains, leading to a decline in the overall amount of energy consumed. In the scenarios developed by the International Energy Agency, for example, such efficiency gains account for more than half the reductions in greenhouse gas emissions by 2035 and nearly 40 percent by 2050. In these scenarios, the total amount of primary energy used worldwide declines as a result of the decrease in the quantities consumed in industrialized countries. This decline is higher in other projections, particularly the studies carried out by Ernst von Weizsäcker at the Wuppertal Institute, based on the idea of an overall fivefold reduction in the resources used per unit of wealth produced.[5]

Simple arithmetic makes it easy to understand the underlying mechanism: if the efficiency gains pertaining to use halve the amount of energy required to obtain the same level of satisfaction, then we can obtain the same service for an energy price that has doubled. In other words, at the macroeconomic level, we will have access to energy services equivalent to an unchanged cost, despite the doubling of the price of primary energy. This equivalence implies that, in what is termed the development of the energy bill, innovations and productivity gains applying to energy services may be separated from those applying to primary energy production. Yet when we carefully examine the past, we realize the importance of innovations that have contributed to the spread of energy efficiency.

The Price of Lighting:
The Proper Use of the "Rebound Effect"

One of the best-documented cases in terms of productivity gains in energy services concerns lighting.[6] The evolution of primary sources used for lighting reflects the historical transformations of the energy system. From the use of candles alone—an indirect use of biomass through tallow, a mixture of animal fats—there were successive shifts to town gas produced from coal, then the oil lamp, and finally the electric light bulb, the use of which spread widely with the extension of the grid, gradually supplanting other forms of lighting from the 1920s onward. Overall, these changes led to a fall by a factor of a hundred in the price of lighting in the twentieth century, thanks to cumulative productivity gains.

The dramatic decline in the cost of lighting may be broken into two factors: the reduction in the price of energy used, itself the result of productivity gains for its generation, and advances in the effectiveness of converting it into light. There is no doubt about the outcome of the process: by and large, we now have access to very cheap lighting thanks to the enormous productivity gains resulting from the general use of more efficient processes in the transformation of primary energy into light. The introduction of and successive improvements in the incandescent light bulb—over more than a century and a half, if one goes back to the original invention[7]— contributed to the bulk of these gains, which have allowed the large-scale dissemination of electric lighting. Such gains, moreover, have by no means come to an end, given the planned withdrawal of filament bulbs, which convert only 5 percent of the energy into light and waste the remaining 95 percent.

Historical observations also provide valuable information on how consumers use these productivity gains. Fouquet and Pearson have painstakingly studied the behavior of consumers in the United Kingdom over a 200-year period.[8] In the early stages of the diffusion of new lighting technologies, consumers were willing to spend a greater portion of their income to improve their lighting and consequently responded to a decrease in the price of lighting with a more than proportionate increase in their purchases. As their standard of living rose, the majority of households became equipped and their sensitivity to lower lighting prices fell. Since 1980, needs have been largely covered, and lighting accounts for only a tiny proportion of the household budget. An increase in people's income now has virtually no impact on consumption, which nevertheless remains negatively correlated with price.

The key idea here is that the impact of energy prices on people's behavior depends on two key parameters: access to services— consumers must have access to electricity in order to benefit from productivity gains brought by the incandescent bulb—and standard of living. In countries with low or intermediate living standards, or within a country for social classes with low income, gains in energy efficiency will therefore mainly result in increased pressure on the demand for energy to cover basic needs. In countries with higher living standards or within a country for the middle or upper classes, such gains will release more purchasing power for other uses. In both cases, these efficiency gains are an irreplaceable transmission belt for the energy transition to generate beneficial economic impacts.

From the standpoint of pressure on resources, environmental economists often deplore the existence of an undesirable "rebound effect" on the consumption of primary energy sources following efficiency gains in use.[9] As we have just seen, the strength of this effect appears to be related to the level of wealth. When basic energy needs are not initially covered, this effect is high and desirable because it can become an instrument for development and action in response to energy precariousness. On the other hand, it is clearly necessary to ensure that the price system leads to increased use of primary sources that are least harmful to the environment and climate. Hence the importance of correctly pricing the different primary sources, starting by including the value we attach to climate stability so as to slow down the release of stocks of fossil carbon into the atmosphere.

The Price of CO_2 and the Uses of Fossil Energy

In the short term, the introduction of a CO_2 price changes the order in which it is advantageous to use different fossil sources. The three primary fossil energy sources by no means have the same impact on global warming: for the same amount of energy delivered, coal emits on average twice as much carbon as natural gas, while petroleum products are situated approximately midway.[10] It immediately follows that the introduction of a carbon price proportionately affects coal twice as much as gas. Whenever uses are substitutable, an increase in the carbon price therefore gives an advantage to natural gas, saving about half the emissions of greenhouse gases into the atmosphere. Within a simple logic of substitution, natural gas can play a beneficial (and transient) role as a transition energy.

Substitution of one fossil fuel for another has so far been the primary driver of emissions reductions following the introduction of a carbon price in Europe through the cap-and-trade scheme. It has taken two main forms: the substitution of gas for coal almost everywhere in Europe and the substitution of standard coal for lignite in Germany and Poland. Tens of millions of tons of CO_2 have been economized by a change of merit order, in which operators use existing plants when they need to increase the load on the network. But these gains obtained from the existing capital stock are reversible. Thus they were lost when, through lack of political determination and with virtual indifference, the European public authority allowed the carbon price to collapse in 2011.[11] To eliminate this risk of reversibility, it is necessary to introduce a carbon price in the form of a rising trajectory over time. This may be done through a reserve price system, as in the case of the Californian allowances system, or by means of a tax whose rate increases over time, as in the case of the carbon tax introduced in France in 2014.[12]

In the medium and long term, the price of carbon should hasten the deployment of technologies enabling fossil fuels to be used without releasing CO_2 into the atmosphere. Let us examine the case of coal. Its proven reserves are in excess of 180 years of production. They are plentiful and more evenly distributed throughout the world than oil. The abundance of coal results in low extraction costs. It is naive to think that a stock of energy so easy to extract will be left in the ground, particularly in emerging markets well endowed with deposits, such as China, India, and South Africa, whose development requires the increasing mobilization of energy sources.

Coal will continue to be used for a long time yet, but it needs to quickly be subject to a simple rule to limit its climatic impact: the systematic recovery and underground storage of CO_2. Such a rule will seem as natural a few decades hence as the existing ban on throwing out wastewater onto the street, a practice that was standard in European and American cities until municipal authorities began, in the nineteenth century, to invest extensively in sewage and water treatment systems. Proportionally, however, the investment needed to capture and bury CO_2 emerging from chimney stacks is much lower. A similar rule should therefore apply to such discharges.

The first experiment in carbon capture and storage on an industrial scale was on the Sleipner gas platform in the North Sea. The gas from the Sleipner field contains 9 to 12 percent CO_2, preventing its commercial exploitation in that state. Since 1996, the Norwegian operating company has recovered most of this CO_2 on the platform and then reinjected it into an aquifer below the seabed. It processes a million tons of CO_2 a year, thus avoiding its emission into the atmosphere. The launch of the operation was facilitated by the existence of a tax in Norway of €40 levied on each ton of CO_2 emitted from oil and gas installations. Reinjection of gas underground thus saves €40 million a year in tax. In these very specific conditions, a carbon price of €40 is enough to make the operation profitable.

Technologies for using fossil fuels by drastically limiting their atmospheric emissions fall into three basic stages: capture, transport, and underground storage of CO_2. There are several competing technologies for capturing CO_2. They constitute an additional investment cost and subsequently weigh on operating costs because they reduce the efficiency of power plants and increase maintenance costs. The transport of CO_2 presents no particular problems, though it makes sense to find underground storage reservoirs close to the capture facilities. There then needs to be agreement on monitoring methods to obviate the risk of CO_2 being released at some future point,[13] a consideration that raises questions as to the division of responsibility. Who will be responsible in 500 years' time for the CO_2 that is injected underground today?

The large-scale deployment of these technologies, each stage of which has already been mastered, comes up against economic barriers, as studies reveal that pilot projects could reach break-even levels with a price per ton of CO_2 of around sixty to ninety euros, significantly higher than the levels currently prevailing in the markets. It is for this reason that additional support has been mobilized in Europe to finance the first industrial pilot schemes. The initial experiments show that in addition to the standard obstacles there is considerable resistance from local residents, who are generally opposed in principle to the transport and storage of CO_2. The price of CO_2 does not allow this question of social acceptability to be answered, just as it is not, in itself, a signal indicating the right choices to be made in terms of the development of biomass energy, for which externalities other than those associated with climate need to be taken into account.

The Development of Biomass Energy

Although the price of CO_2 measures the actual importance we attach to the stability of the climate, it is not a general standard capable of measuring the cost of all the damage adversely affecting natural capital. The choice between gasoline or diesel engines, decisions concerning the use of biomass and the development of biofuels, along with many other examples, point to the need to price all externalities.

In the existing energy system, biomass is the primary source used after coal, oil, and gas. It is a renewable resource when it is taken from a stock that can reproduce itself. Otherwise, its use may contribute to the destruction of the natural environment. There are many types of biomass use, ranging from the burning of wood for fuel—its most common usage—to its transformation into biogas or biofuel. Many other materials may also be used for energy production, including wood, dedicated crops, byproducts, marine algae, waste, etc.

The energy-producing potential of biomass as a substitute for fossil fuel constitutes one of the important options in the energy transition. In the IEA's low-carbon scenario, the development of biomass as fuel halts its secular decline in the energy mix. IEA experts even suggest that from 2040 biomass power plants will be producing electricity and then capturing and storing the CO_2, thereby making them "negative net emissions" energy producers: the more energy they produce, the more CO_2 they remove from the atmosphere.

If, however, the value of the climate is the only externality taken into account, there is a risk of placing too great a burden on the natural environment and incurring disappointments, as happened in Europe with first-generation biofuels. It was initially decided to aim for an incorporation rate of 10 percent biofuels to help attain the overall goal of 20 percent renewable energy by 2020. Upon implementation, it became apparent that this target was liable to have indirect effects on land use change and in particular might increase the rate of deforestation in the tropics. Hence the addition of various, more or less bureaucratic, criteria to distinguish the "good" biofuels allowed in the European market from those whose development contributes to tropical deforestation and loss of biodiversity.

As Gabriela Simonet shows,[14] using the sole criterion of CO_2 emissions reduction may also orient forestry projects toward fast-growing, frequently monoculture plantations that optimize energy efficiency at the expense of the diversity of ecosystems. Such is often the case with the types of projects developed in China with the help of the World Bank and the financing of carbon credits acquired in the framework of the Kyoto Protocol mechanisms. These choices would not have been made if a value attached to biodiversity had been included in the pricing systems.

What Simonet perceptively investigated in the context of development projects also applies in developed countries. If one wants to quickly generate the greatest amount of biomass energy as an alternative to fossil fuels, one should not count too much on the giant sequoias of California or the ancient oak forest of Tronçais in France, some of whose living specimens were planted by the French statesman Jean-Baptiste Colbert in the seventeenth century. In the name of the energy transition, are we prepared to chop down these forests and plant short-rotation coppices, which, though ecologically impoverished, are much more efficient from the energy and climate angle?

Distributive Issues of Energy Pricing in the Transition Strategy

Energy prices are often presented in the public debate as a set of unpredictable parameters, appearing in opaque and sometimes manipulated international markets to which governments have to submit. This view, which absolves policy makers from accountability, must be rejected. To involve the energy system in the ecological transition, it is essential to introduce the costs associated with climate protection and more generally with the reproduction of natural capital into the pricing of energy.

Such pricing has the effect of raising the cost of most of the energy sources currently used. In the medium term and well beyond these increases can be absorbed by the speed-up of gains in energy efficiency, one of the keys to the transition. But it takes time for these gains to be diffused in society, especially among the disadvantaged social classes, who are likely to shoulder the burden of the process unless affirmative measures are taken simultaneously.

In this kind of situation, it is very tempting to engage in price-distorting action. This may take the form of a general subsidy promoting the use of certain types of energy, as in Iran, the majority of Gulf countries, and Indonesia for fossil fuels. This type of support, which is very expensive for public finances, is virtually no longer used in developed countries. But production subsidies are still commonly used, as in Germany for its coal mines or indeed in Ireland for the extraction of peat. Financial support for professional use of certain fuels—diesel in most cases—is still the norm in Europe. And the fight against energy precariousness often involves the use of social tariffs subsidizing electricity and gas. In France, according to the Energy Mediator, nearly one million households are affected.

The widespread use of these tools is a major obstacle to the financing of the energy transition. When all the subsidies are aggregated, it is clear that they far exceed the resources that governments are able to bring into alignment to support the development of new energy industries. Reducing support for fossil fuels would thus free up precious budgetary resources for other energy uses.

At the microeconomic level, such support has the primary effect of locking households and the relevant professions into systems encouraging the overconsumption of fossil fuels, while making them even more vulnerable to inflation in the future. To establish a sound basis for the energy transition, it is important to find new forms of support and solidarity that do not distort the signals that energy prices should be sending to society as a whole. This transparency of prices and costs should also apply in a sector in which it has long been absent in France: nuclear energy.

16

Nuclear Energy

A Rising-Cost Technology

THE PHENOMENON OF NUCLEAR fission was first described by the German chemist Otto Hahn in December 1938. It resulted in accelerated military developments and the dropping of the two bombs that hastened Japan's surrender in 1945. Civilian applications were equally rapid, with some observers in the 1970s detecting the beginnings of a true energy transition thanks to this new source, which at the time seemed to offer a bright future.

A nuclear power plant uses energy from fission, the chemical chain reaction caused by splitting the nuclei of uranium atoms. The energy produced is then converted into electricity by means of a turbine. This process releases no greenhouse gas emissions. All other things being equal, a CO_2 price thus tends to decrease the cost of nuclear power compared to fossil fuel energy. Would the widespread introduction of carbon pricing then justify the use of nuclear power, in the name of the transition to a low-carbon energy system?

As astutely analyzed by Ulrich Beck in *Risk Society*, public opinion reacts much more according to its perception of new risks associated with the nuclear industry. Three major accidents involving nuclear power have given rise to this perception. Nuclear fission takes place in the core of the plant, the reactor, which has to be continuously cooled. Any failure in the cooling system exposes the reactors to the risk of meltdown of the fuel, which may then escape from its containing vessel into the environment, along with all its radioactive material. Accidents of this kind have occurred on three occasions in the history of the industry: at Three Mile Island in Pennsylvania in 1979, at Chernobyl in 1986, and at Fukushima in 2011. The price of CO_2 clearly does not take into account this type of externality. How then should the debate on the future of nuclear power be framed?

The Fukushima Shock Wave

For critics of nuclear power, the Fukushima accident was but the latest illustration, fifty years after the first plants were commissioned, of the extent of the risks associated with nuclear energy. Fukushima shows, firstly, that not all combinations of circumstances can ever be anticipated: the Japanese power plant was subjected to an earthquake of unprecedented strength and a tsunami that rendered the cooling system inoperative. In addition to these natural hazards, there were human errors at the time of the accident and subsequently in terms of crisis management. These natural and human risks surrounding nuclear power come on top of those related to the processing and storage of radioactive waste. There is also the threat that fissile material may be diverted from civil applications and used to make nuclear weapons. The only reasonable option, the critics say, is the phasing out of nuclear power.

Germany, for example, made this choice following the Fukushima accident. Echoing the profound concern of a public largely hostile to nuclear energy, Chancellor Angela Merkel, who had originally been favorable toward the industry for reasons of economic realism, had a change of heart and adopted a strategy of total exit by 2022.[1]

The supporters of nuclear power can no longer deny the risks, but they interpret the situation differently. First they acknowledge that the Japanese accident should lead to a re-evaluation of the risks. They also point out that all energy systems entail risks and environmental costs. How many people die every year in coal mines or from local pollution caused by the use of coal? How about the environmental cost of oil spills? Are the sometimes very toxic rare metals used in solar panels correctly recycled? It is therefore necessary—though by what method?—to compare the respective risks of the different energy industries before reaching a conclusion.

The main argument put forward by proponents of nuclear energy, at least in France, is economic in nature: it enables non-carbon-based electricity to be generated at a moderate cost. Abandoning it would lead to increased electricity prices and thus penalize competitiveness. In practical terms, this argument is strengthened every time a CO_2 price necessary for the preservation of the climate is introduced. Moreover, up until Fukushima, taking the climate constraint into account seemed to have given new life to the nuclear industry. As its lobby does not hesitate to point out, electricity production is the world's primary source of greenhouse gas emissions, and most new production facilities constructed in emerging countries use coal. Is nuclear power an alternative to coal?

Nuclear Energy Will Remain a Backup Source

The best way to disentangle these conflicting arguments is to start with the facts. In 2012, thirty of the world's countries were operating at least one plant, or just under one in every six countries. Civil nuclear power is an industry of the high-income countries and only started spreading to emerging countries in the late 1990s. Among the developing countries, only China has managed to join the club of the top ten producers, which control 85 percent of the world supply. It is quickly catching up, with the largest number of new plants under construction and rapid acquisition of the relevant technologies. India may soon follow. If the transition to nuclear power is inevitable, it must be recognized that it is not accessible to all countries, as the acquisition costs of the technology and the tying up of capital are considerable barriers to entry. For a form of energy that can be made widely available, particularly in the developing world, there are better alternatives than investing in nuclear power.

But can nuclear energy facilitate the path to a low-carbon energy system in the world's high-income countries? Here again, it is instructive to look at practical experience. In Germany, despite considerable investment in renewable energy, energy efficiency, and distribution networks, there will need to be a temporary increase in the proportion of electricity generated from gas and coal to the detriment of the climate challenge. Moreover, this strategy is fundable only through a fiscal effort that no other major European country would be able to make and by imposing energy prices on households that are among the highest in Europe.[2]

For a country whose nuclear power plants are partially amortized, rapid exit from nuclear energy therefore imposes additional transition costs, both economic and climate related. Hence an intermediate position involving a compromise strategy: extend the use of existing power plants, but without building any new ones. In the United States, for instance, where such a strategy has been adopted, no new plants have been commissioned since 1979, the year of the Three Mile Island accident. Attempts by President George W. Bush to help revive the industry ended in failure. It is too early to assess President Obama's policy of stabilizing the regulatory framework and providing massive federal government guarantees for private investors prepared to take the plunge. One thing is certain: experience shows that it is very difficult to attract sufficient private capital to this industry deemed at risk. There are cases in which Anglo-Saxon capitalism may foresee the health and environmental risks more clearly than might be supposed: American private insurers stopped covering the risk of exposure to asbestos back in the 1920s, whereas the French government waited until 1997 to completely prohibit its use.

With a few exceptions, including France, the debate over whether to focus on nuclear power as a major source of electricity thus seems outmoded. The question arises differently in emerging countries facing a rapid increase in the demand for electricity, where nuclear power will continue to be called on to reduce pressure on the use of coal and gas. But even in the most favorable scenarios, nuclear power will play only a backup role in the energy transition: it now accounts for 6 percent of the primary energy used in the world and 14 percent of electricity production. In the International Energy Agency's most positive scenarios, developed before the Fukushima shock, it will account for just 10 percent of primary energy and 19 percent of electricity production in 2030. Since Fukushima, the consensus has drastically revised those figures downward.

Lessons from the French Experience

France is distinctive in having a nuclear energy capacity that provides nearly 80 percent of its electricity and has given rise to industrial export champions. The development of this sector is a rare example of a successful Colbertist (state interventionist) industrial policy, consistently followed by successive governments over several decades, with the active support of the major trade unions. President Hollande is the first to have dented this tradition, in promising during his election campaign in 2012, then deciding once elected, to reduce the share of electricity generated from nuclear power to 50 percent by 2025. This political decision, taken before the start of the debate on energy transition in 2013, was the result of a rather astute compromise with the environmentalists: by closing Fessenheim, the oldest French nuclear power plant, no further choices need to be made by the end of the presidential term, in view of the age pyramid of the plants. But the country will one day have to decide whether to continue investing in this energy sector, to which a considerable proportion of public resources are directed, including research and development, and are therefore not available for renewables or energy efficiency.

This first step in the diversification of the electricity supply offers many benefits and will probably be taken further in the coming years if France wants to make a success of its energy transition and to redirect, under good economic and social conditions, the resources currently tied up in nuclear power. On examination, the arguments in favor of nuclear are, and will be, less and less convincing, whether they involve economic Colbertism (national champions), a social variant of this (cheap and abundant electricity in France), or climate considerations (acceptance of radioactive hazards in the name of saving the planet).

Much of the public effort made in the industry was prompted by the objective of building a strong position in a seemingly promising global equipment and services market. The argument was based on the assumption of a rapidly expanding nuclear industry outside France, an expectation that seems increasingly at odds with reality. The story begins in a way rather reminiscent of Concorde, a Franco-British technical achievement that came to a sorry economic end. Instead of the nuclear industry, the most promising energy markets internationally seem to be in renewables, energy efficiency, and, indeed, the new niche of end-of-life decommissioning of nuclear power plants. Experience has shown a recurring tendency at the time of their launch to underestimate the time and cost of building new plants. And so far nobody really knows the time and cost involved in dismantling them and processing the waste at their end of life.

The claim that nuclear power provides base-load electricity at a competitive cost has long been difficult to substantiate because the real costs of nuclear power in France have remained opaque. Most opportunely, considerable light was shed by a report from the national Audit Court published in 2012.[3] It appears that the cost of nuclear electricity tends to increase over time, mainly due to better recognition of the associated risks—a state of affairs that is rather good news from the security standpoint. There are, in addition, the decisions made regarding the technologies adopted in third-generation plants. As a direct consequence, the marginal cost of a kilowatt hour produced by EDF at the third-generation plant in Flamanville in Normandy is likely to be higher than the base price of electricity purchased by households, a situation that is very worrying for a power plant intended to cover basic needs at minimum cost.[4]

The advocates of nuclear power will retort that the cause of climate change is likely to completely reshuffle the deck: with a good price for CO_2, won't the competitiveness of French nuclear power be finally and permanently restored? That is almost certainly the case if the carbon price is high enough, but only compared to fossil sources. The real issue for the future is much more that of anticipating the capacity of nuclear power to maintain competitiveness in the face of renewables and other technical solutions that will allow massive energy savings to be made, and will allow the CO_2 from the combustion of fossil fuel to be captured and thus prevented from accumulating in the atmosphere.

In this regard, the debates among experts may continue for a long time yet. The historical trend, however, is in little doubt, as François Lévêque points out in one of the most comprehensive books on the subject:[5] since its inception, the production cost of a kilowatt hour from nuclear energy has been increasing, while costs of renewables and other low-carbon innovations have been steadily falling. Except for onshore wind, the costs of other low-carbon technologies are still generally higher than those of nuclear power. But for how long? If you were a private investor, into which industry would you put your savings within a medium- and long-term perspective?

Given the respective risks of nuclear technology and of climate change, is it a choice between Scylla and Charybdis? Resolving this dilemma entails mobilizing all the resources needed to develop alternative technologies that economize on energy or use it without releasing CO_2. Until such time as this happens, nuclear power represents not an alternative to fossil fuels and renewables, but a backup option to be used with caution. France, having gone further down the nuclear road than any other country, has been able to build one of the world's least CO_2-emitting electricity systems. Given the re-evaluation of risks associated with this technology, France must nevertheless be prepared to accelerate the diversification of its energy mix so as to overcome its excessive dependence on nuclear power—probably an "energy of the future that belongs the past," in the well known phrase of the American essayist Amory Lovins.[6]

17

Growth-Generating Innovations

IF, IN 1900, ONE had questioned the most perceptive economists about the technical changes that would structure the economy of the coming century, none would have been able to point to the innovation clusters of the late nineteenth century, now considered by historians to be crucial to growth. These innovations were already there, but no one could have anticipated the complex interactions that would make them the real drivers of the expansion of the twentieth century as a result of their widespread dissemination. A prototype internal combustion engine was certainly presented at the World Exhibition in Paris in 1900, but its inventor, the engineer Rudolf Diesel, was viewed as a marginal figure, passionate about his wheeled vehicles powered by peanut oil. At that time, Thomas Edison had filed most of his thousand patents, including those for the filament bulb and the first electric power plant, inventions that would transform people's lifestyles; yet he was regarded in Europe as a former Western Union telegrapher, adept at purloining the inventions of others, particularly the early films of Georges Méliès, which were used in questionably legal circumstances to develop Kinetoscope parlors, the forerunners of cinemas.[1]

Nearer to the present day, in 1980 Bill Gates signed a cooperation agreement with IBM for the commercial development of the MS-DOS operating system that would make his fortune. Would any economist then have been able to predict the changes triggered in the functioning of societies, not by computers, machines that IBM had been perfecting for decades so as to maintain its dominant position, but by their networking through the Internet? The first applications of the Internet had already been in operation for several years among academic centers in the United States, but no one foresaw that this type of application would one day emerge from the very restricted environment of American research and become accessible in a few clicks on today's cell phones.

The experiences of our predecessors show that we should proceed with extreme caution when it comes to predicting the future, as there is no reason why we should be any more far-sighted than they were. The technological innovations that will underpin the fifth energy transition are probably already largely in place. But it is impossible to identify the combinations that will produce the economic and social changes enabling the shift to a low-carbon system. Nor do policy makers possess the gift of clairvoyance any more than economists. They therefore have to find the right system of incentives facilitating the deployment of these innovations in the economy without favoring any particular technological option, within which the various actors could become imprisoned. The construction of possible scenarios, revealing the set of constraints and interdependencies among the multiple links of the energy system, can be of help here. Such scenarios also clarify the normative objectives to be targeted—in other words, the kind of future we want to build. One thing is certain: nobody can say what the energy transition will really be like.

The Shift from Stock Energy to Flow Energy

In the current vision of energy futures, breakthrough innovations are often presented as new, preferably spectacular, vectors enabling the secular constraint of energy scarcity to be suddenly removed: plants or marine algae with miraculous yields that greatly increase access to biomass; the discovery of a hyperconcentrated energy source brought from another planet; exploitation of the enormous power of the winds blowing tens of kilometers above our heads; the long-promised mastery of nuclear fusion, finally providing electricity in vast quantities and at low cost. . . .

These futuristic Jules Vernian reveries take us far away from the main characteristic of the fifth energy transition, which lies precisely in the extraordinary diversification of sources and methods likely to ensure tomorrow's supplies. Such diversification will be possible through incremental improvements in the production techniques of flow energies (wind, solar), whose costs are currently approaching, or even outperforming, those of the stock energies (fossil fuels) still dominant today.

With regard to fossil fuels, financial investment in research and exploration continues to grow. Sometimes these investments produce dramatic results, as with the exploitation of shale gas and unconventional oil in the United States, which for the next few decades will ease tensions around fossil resources. But the underlying trend remains: the physical return from these investments—the number of tons of oil equivalent obtained per dollar invested—is decreasing over time, and this in turn raises the cost of these energy sources. Furthermore, technological advances in the transformation of these primary sources into final energy do not always follow. Though technological progress continues to be made in the propulsion of transport systems—land vehicles, aircraft, and ships—it is stagnating in electricity generation: the energy yield of thermal power plants has been at a virtual standstill since the end of the 1960s.

With regard to flow energies, the amount invested has been much smaller, as the carbon price has not sufficiently changed compared to fossil fuels. In some cases, they are still only at the laboratory experimental stage. The two flow energies expected to play a major role in the transition are wind and solar, the only renewable energies in the truest sense of the word.

Wind power combines the ancient technique of the windmill with a process for converting mechanical energy into electricity that has existed for more than a century. The economics of wind power production is quite simple: it is based on first choosing the right windy location and then optimizing the energy performance, which is dependent on the diameter of the blades, itself related to the size of the tower. The first wind turbines were less than 15 meters in height, with a power output of around half a megawatt. Recent generations measure 125 meters and have a capacity of about 5 megawatts. The sector's manufacturers foresee turbines of up to 250 meters in height, with a power output of around 20 megawatts, in future offshore wind farms. This increase in the size of the equipment has been the main factor in reducing the costs of an industry that shares two features with solar: it is wholly capital intensive, as it operates without production manpower or intermediate consumption, and it is dispersed over a large number of production sites. Its production costs are or will be lower than thermal electricity. But this upstream competitiveness is only one of the conditions of its deployment in the energy system.

The solar industry converts the energy in the sun's radiation striking the planet by means of two methods. The thermal process involves directly using the heat present in this radiation, for example to heat domestic water or to vaporize water in the turbines of solar power stations by means of reflectors. The photovoltaic process converts solar energy into electricity through panels comprised of photovoltaic cells. The thermal process for domestic use is now often competitive compared to alternative systems. Photovoltaic is benefitting from very rapid productivity gains in the manufacture of silicon cells upstream of the industry, though this only represents part of the cost of installations. Unlike the solar water heating segment, its production costs, although decreasing, are still higher than those of conventional thermal sources in most applications.

The standard argument as to the high costs of producing flow energies thus must be evaluated on a case-by-case basis. In trend terms, these costs are tending to decline as a result of increased research and development efforts and the many experiments that constitute the usual learning curve of economic sectors in the emerging phase. Carbon pricing, one of the conditions for the large-scale deployment of these new sources, will speed up this scissors effect. As occurred with lighting, these incremental advances in technology could lead to real breakthrough innovations if a profound reconfiguration of energy networks enables the widespread dissemination of their uses.

The Reconfiguration of Energy Networks

For some decades now, energy systems have not reduced their dependence on fossil fuels, but instead have tended to organize themselves in a "silo" pattern. Electricity accounts for an ever-increasing proportion of final consumption, and its production has become the main outlet for coal and to a lesser extent gas. Transport and distribution infrastructures are thus constituted as a vertical network, passing through power plants whose size has tended to grow over time. A parallel silo of less importance (except in France) has been built for the nuclear industry. At the same time, transportation has become the main market for oil and its derivatives. With the exception of passenger trains and trams, almost all forms of transport use petroleum as the primary energy source. Hence the third infrastructure silo, linking transportation systems to oil wells situated immediately upstream of the industry.

Within each of these silos, economic actors dominated by large energy companies have worked to optimize and secure supplies, with some success, as the cost of energy made available to the user in recent decades has grown more slowly than wages or the general price level.

The most obvious effect of this configuration of the energy infrastructure into vertical silos is to drastically reduce the possible choices of energy consumers and to loftily ignore the possible complementarities among sources and uses on a territorial scale. In this context, the introduction of still very tentative flow energies is implemented through special systems (that are rather expensive for local authorities) or limited to small-scale experiments in the case of transport (e.g., the sharing of electric vehicles in cities). These transient models allow only limited amounts of flow energy to be introduced. The large-scale deployment of these energies requires a drastic reconfiguration of networks that will break down the vertical silos and the various economic interests and dominant positions attached to them. Such a reconfiguration will fundamentally change the way energy is used in the future by multiplying the range of sources available and eventually erasing the distinction between producers and consumers. It will affect two key areas of people's lives: housing and transport.[2]

Habitat accounts for between 30 and 40 percent of primary energy consumption (heating, but also specific consumption related to cooking, lighting, and the use of various appliances). For nearly all appliances, there are potential energy efficiency gains as large as those highlighted for light bulbs. Furthermore, better programming of their use would eliminate demand peaks, thus constituting a major source of efficiency. It also important for producers to be given incentives by means of technical standards and pricing by use and not by product. With regard to new housing, architects now know how to design positive energy houses, in which more energy can be produced than consumed if their functionality is optimized. Advances in insulation techniques would often give comparable performance in older housing, but at rapidly increasing costs depending on the initial quality of the buildings. In both cases, optimizing energy use is based on proven technologies or those currently being developed. What is missing is a system of incentives to ensure their wide dissemination. This last point is the main issue in the reconfiguration of energy networks.

Strangely, in today's digital age, energy networks have remained in the Stone Age. Their reconfiguration will play a pivotal role in the energy transition. The interconnection of smart meters would allow millions of users to manage energy flows entering and leaving their homes by taking into account information about their own consumption and incorporating this into information coming from other users. To send the right signals to users, this physical infrastructure should be used to support pricing that reflects the relative scarcity of different energy sources at all times. Such emergence of multipolar interactive networks is the antithesis of centralized top-down networks, the archetype of which is probably the French electricity grid, built for the distribution of a product as inflexible as might be expected given that four-fifths of the country's generating plants are nuclear.

With regard to transport, it seems that society has in recent decades wanted to travel ever faster, a race that both fascinates and disturbs us: the critics of Rudolf Diesel at the time claimed that the human body could not tolerate a speed of 100 kilometers per hour. For many reasons, this quest for speed seems absurd. Urban freeways are clogged with superbly engineered, oversized vehicles; airliners reduce their cruising speed to lower their operating costs but waste fuel waiting for their landing slot due to traffic congestion; at rush hour, passengers hurry down metro passageways at the sound of an approaching train, as if avoiding three minutes standing on the platform were a vital issue. The reconfiguration of transport networks calls for preliminary thinking about what types of mobility of people and goods society needs.

In this way the emphasis shifts from a product economy—cars, trains, aircraft—to a functionality economy.

In such an approach, the benefits of mobility will be multiplied if the modes of transport for different requirements, as well as the energy sources used, are diversified. Regarding energy, such diversification will give a second life to certain age-old sources, such as human muscle power with the rediscovery of the benefits of cycling and walking, and will generate many new forms of hybridization between modes and between energy sources. With these changes, the car will not disappear but will be integrated into broader mobility patterns. The car's propulsion will call on various sources: conventional fuels; biogas and biofuels; hydrogen and electricity. From an energy perspective, the ramp-up of vehicle electrification will realize the huge potential for decentralized electricity storage in vehicle batteries, including battery charging during off-peak periods. The general adoption of this type of behavior is likely to generate significant energy and greenhouse gas emissions savings through tomorrow's decentralized open networks.

The augmentation of our still very limited electricity storage capacity should be a major parameter in the reconfiguration of energy networks, which will help overcome in particular one of the most constraining current limitations in the distribution of energy flows: their variability. Another constraint is the low density of these energies in an environment where populations that are ever more densely concentrated in cities need to be supplied. It is for this reason that the energy transition will require rethinking the links between the energy system and territories.

Mastering Space: The New Territorial Issues

The historical transitions of the past led to the use of increasingly concentrated energy sources: a ton of oil contains more energy than a ton of coal, which in turn contains much more than a ton of wood. This concentration greatly facilitates transport and storage and limits the area used by energy systems. From this standpoint, the shift from animal to motorized traction freed up large areas of farmland. For example, in 1955, a quarter of the French grain-producing area was still tied up in growing oats for feeding draft animals. This crop has virtually disappeared today and, despite its rapid growth, the area used to grow biofuels is no more than 5 percent of the total farmed area.

A notable feature of the fifth energy transition is that it relies on the transmission of new energy flows and of renewable biomass, which tend to be geographically dispersed.[3] That is why they are easier to deploy for meeting the needs of scattered populations rather than large cities, which are, moreover, trying to densify their networks. Solar is now the most competitive source for operating pumps bringing water up from rural wells located far away from electricity grids.[4] It can be selectively used in rural houses to pump previously collected rainwater—though that requires additional financial input—but it can only marginally contribute to the provision of energy needed for supplying and processing drinking water for cities. Similarly, it is technically much easier to design and build a positive energy house (especially if it has enough land) than a positive energy apartment building.

This spatial dimension should not be underestimated. Despite its concentration, infrastructure installed for the extraction and distribution of fossil fuels occupies about 30,000 square kilometers worldwide, an area the size of Belgium. Ignoring space constraints in the development of new energy sources with lower density can lead to major disappointments, as revealed by the experience of the first generation of biofuels. Technologies for using energy from byproducts and waste (biogas, conversion of woody materials, etc.) or new sources (algae) will help ease the spatial issues related to the use of biomass. But these issues could soon recur if we erect wind turbines and install photovoltaic arrays in the countryside to supply cities with large amounts of energy.

In terms of governance, this state of affairs implies that local authorities should have increased room for maneuver, responsibilities, and powers. One of the major difficulties in creating decentralized governance is to couple it with new instruments for ensuring solidarity among territories. In a centralized network, the key instrument of such solidarity is the equalization of pricing. In tomorrow's multipolar and open networks, it will be necessary to find other instruments that no longer hamper the role of prices in the proper allocation of resources.

Time Management: How to Leapfrog Stages

If space is limited, time, for its part, is running out! Smil (2010) has pointed out how past energy transitions involved secular transformations, with considerable intervals between the times when innovations appear, when they spread in society, and when they exert their boosting effect on the economy through productivity gains. Technological constraints, the inertia of economic and social structures, and the power of vested interests all account for the slow pace of energy transitions. History also reveals a surprising paradox: emergency situations, instead of hastening the transition, frequently lead to resorting to old solutions that impede changes in the energy system.

Smil gives the example of the celebrated Liberty ships hurriedly ordered by the U.S. Navy after the attack on Pearl Harbor: built in record time, 2,710 troop carriers emerged from American and Canadian shipyards between 1942 and 1945. To speed up the process, the propulsion system chosen did not use diesel engines, even though they had been introduced on transatlantic shipping lines from 1911 onward and had equipped the majority of new vessels since the late 1920s, but instead used ancient steam engines, which were less efficient and more polluting. The decisive advantage of this proven but outmoded technology was that its robustness reduced the risk of late delivery and thereby answered to the urgency of the situation.

A similar urgency argument is currently used to justify opting for fossil fuels, as seen in particular in the frantic pace of construction of thermal power plants in emerging Asian countries. In Mumbai, the economic capital of India, the population is growing by more than 5 percent per year as a result of the rural exodus. A quarter of the population is crowded into shantytowns such as Dharavi, home to nearly a million people. Giving them access to decent housing connected to electricity and a sewage system has to be the number one priority. The fastest option to answer this need often seems to be the hasty building of coal-fired mega-power plants, of a kind with which India is already liberally endowed. Time is too short to design and implement alternative solutions. As a result, they are reproducing an energy infrastructure system inherited from the last century, whose imprint will shape the energy systems of emerging countries for decades to come.

One of the major issues of the fifth energy transition is to make the urgency of improving the energy supplies in the countries of the South compatible with considerations of global warming. According to the standard interpretive frameworks of development economists, such a reconciliation seems impossible in the short term because these countries have little room to maneuver due to the scarcity of their resources and their technological backwardness.

This diagnosis may in fact be premature. Developing countries are much less constrained than the old industrial nations by heavy energy infrastructure, extensively deployed to support the production and transport of fossil fuels. A larger proportion of the population lives in rural areas, often without access to electricity or other modern forms of energy. As has happened with mobile telephony, the spread of new flow energies will widen access while saving on the costly investment required for deploying the energy networks of the past. Similarly, new models of urban management can develop owing to rapid urbanization, for which solutions directly inspired by urban planning methods developed in the North are not suitable.

We should not think of the next energy transition as a change that will be instituted in the North and then applied to the South. The ramp-up of flow energies in emerging markets provides an opportunity to build new energy models directly, without going through all the stages that currently tie the industrialized countries into the use of fossil fuels. Environmentally, the pace of the transition speeds up whenever these emerging countries produce innovations that allow them to move directly to flow energies without going through the intermediary stage of fossil fuels.

This point very much leads to the issue of the protection of biodiversity. Primary forests are one of the largest reserves of the diversity of living things. Such old-growth forests have been largely lost in the high-income countries and are experiencing multiple pressures with the development of the countries of the South. Like climate protection, the protection of biodiversity also means that developing countries are finding their own paths that avoid reproducing the trajectories historically used by the old industrialized countries.

18

Planning or the Market

What Are the Catalysts?

IN NAIVE OR RIGHT-MINDED opinion, the ecological transition will take place spontaneously, as society suddenly realizes that there exists a common interest in protecting natural capital and leaving a living planet for future generations. The economic approach has shown that the start of this transition depends on the introduction of powerful incentives disrupting the system of prices and revenues so that people attach real value to natural capital and invest in its protection. But how are such incentives to be introduced in the real world where numerous vested interests, threatened by this process, will join forces and the potential beneficiaries, who are often unaware of each other initially, are much less assertive or indeed not yet born? Should we count on the market, reintroduce planning, or mix the two?

Some believe that the market will eventually take care of the environment spontaneously—in which case, why impose additional constraints? When the environmental problem weighs more on growth, the market will take account of it naturally and bring about a transition with all its characteristic flexibility. This type of approach views the market as an autonomous entity that makes its own decisions and achieves in the most natural way the best possible allocation of resources. The proponents of the Chicago school have taken this free market approach to the limit. But dogmatism in this area ends up rebounding on its advocates. Staking everything on spontaneous reactions of the market to get the ecological transition under way is about as effectual as strictly following the prescription of a well known joke in the small world of economists: "The bulb in Professor Friedman's office has burnt out?[1] No need to call the maintenance department, the market will take care of it much more efficiently!"

As a counterpoint to the neoliberal view, permanent mistrust of the market can lead to advocating the setting of prices and quantities by a public authority. In its purest form, this view results in introducing an agent external to the market, a kind of extraterrestrial referred to by economists as a "benevolent planner," whose task involves standing in for the market by establishing a schedule of prices and quantities that takes people's welfare into account. In some countries, at a certain historical moment, the extraterrestrial was called the Politburo—with the economic and environmental consequences we know all too well.

The catalysts for the environmental transition do not lie within either of these two approaches. Instead they are based on the introduction of new environmental values that will generate income transfers, thereby providing opportunities to improve the distribution of wealth through the creation of social acceptance. Situated within a circular economy, they will involve a re-allocation of public investment associated with new forms of governance focused on spatial planning.

Some Lessons from Economic Theory

The standard economic approach is to observe the behavior of agents in their usual environment. Referred to as "laissez-faire" or "business as usual," it involves determining the outcome of these behaviors in terms of consumption, capital accumulation, and income redistribution and the various externalities resulting from them. The "first welfare theorem," used in chapter 7, shows that if markets are perfectly competitive, their overall equilibrium optimizes the welfare of economic agents—the so-called Pareto optimum. Depending on the initial allocation of resources among actors, this equilibrium may be socially acceptable or wholly unjust. In both cases, the theorem says only that this equilibrium is optimal, in the sense that one cannot increase the welfare of an agent without reducing the welfare of at least one other agent.

The probability that this optimum spontaneously reflects the goals that society sets in terms of fairness or environmental protection is infinitesimal. The theorist thus invents a benevolent, centralizing planner . . . in other words an extraterrestrial able to identify all the externalities generated by behavior under business as usual. This Übermensch then calculates an optimal situation in the light of the criteria adopted: social equity, environmental protection, optimized growth, etc. If there is a discrepancy between this theoretical optimum and the situation stemming from business as usual, then the question arises as to how the gap can be closed. What should be done to make economic agents change their decisions? Can one decentralize the optimal solution and, if so, with what public policy instruments?

In the economic literature, the "second theorem of welfare," often less known to students than the first, is supposed to answer these questions. It sets out the conditions under which one can aim at an optimal state of the economy compatible with certain fairness objectives with regard to the distribution of wealth defined outside the market. Its main conclusion is that any Pareto optimum can be decentralized if lump sum transfers between agents are implemented without distorting the price system. In other words, a Pareto optimum allocation of resources can be found through the market following an initial intervention to effect a redistribution of agents' allocations.

This theorem is generally interpreted as a guide for public policy because it is possible to obtain an efficient allocation of resources in the economy through the market, consistent with a redistribution of wealth determined by considerations external to the market. From it we can draw a key lesson regarding the environment: when introducing the value of the environment through pricing, it is essential to clearly distinguish between the allocative effects of prices and their distributional effects. The catalysts we have identified are aimed at getting the ecological transition under way while keeping to these basic economic principles.

The Proper Use of the Price Signal

To be effective in the Pareto sense, the price signal must have two characteristics: it must set the price for the total damage to the environment, and in the same way it should apply to all entities causing pollution. In such cases, economists say that the internalization of the costs of pollution does not cause any "distortion" in the price system. Indeed quite the reverse, for it remedies a major imperfection of the market by improving its allocative function.

But what exactly is meant by "distortion"? Let us illustrate the concept using a concrete example.[2] A company manufacturing metal carts located in Normandy naturally purchases its coils of steel wire as cheaply as possible. A potential supplier located less than 50 kilometers away charges a higher price than a competitor based in Turin, Italy. The cart manufacturer therefore sources the wire from the Italian company. The irony of the story is that the supplier to the company of Turin is none other than the potential supplier of raw wire located within 50 kilometers of the factory— a detour of some 3,000 kilometers to end up with wire produced just over 50 kilometers away! The explanation lies in the system of Italian and European export subsidies for processing wire that distorts the price system.

Among price-based public incentives, those relating to electricity are particularly numerous and have given rise to an extensive literature. Moreover, they have recently been extended in the interests of the objectives of the energy transition. While some of these incentives meet the criterion of economic efficiency, others diverge from it by distorting prices.

Because electricity is still very difficult to store, a peak-period kilowatt hour is considerably more expensive than a kilowatt hour in off-peak periods, as may be seen every day in the wholesale markets. It is therefore consistent with economic efficiency to charge the final consumer less during off-peak periods than at peak times, although having two prices for the same good may offend common sense. In point of fact, a peak period kilowatt hour and an off-peak period kilowatt hour are two totally different products from an economic standpoint. Provided it is unique and applies all the producers, as in the market for CO_2 allowances, a carbon price incorporates environmental pollution by penalizing the most carbon-emitting power plants. It in no way affects the efficiency of resource allocation in the market by prices in the short term. On the other hand, the method used to allocate allowances has significant distributional effects between companies, and it is liable to cause distortions if it provides some producers with unwarranted rents.

Is it effective to introduce an artificial feed-in tariff, higher than market price, to support producers of electricity from renewable sources? The effectiveness of this form of support, practiced on a large scale in Europe, is highly doubtful, as it introduces a number of distortions into the price system: it distributes rents, sometimes very generous, to some of the producers; it increases the price of the good whose distribution one nevertheless wants to promote; if it is the state that makes up the difference, a financially ruinous mechanism is established; and if it is consumers as a whole who share the bill, the result is generally anti-redistributionist because the primary users of renewable energy typically belong to the wealthier classes. So should "social tariffs" be instituted to combat energy poverty? At first glance this seems like a good idea because it allows the energy bill to be shared among rich and poor. However, such social tariffs give disadvantaged households an undesirable incentive to overconsume energy. That is why it is more consistent with the second theorem of welfare to fight against energy poverty by initiating income redistribution and/ or proactive social and professional integration policies without distorting the price system.

The primary catalyst for the ecological transition is the introduction of environmental pricing for remedying a major market imperfection by internalizing environmental values without distorting the price system. And that covers it for the allocative function. If the operation is carried out on a sufficient scale, it creates a new value, "environmental rent," which raises new challenges in terms of the distribution of wealth in society. We now turn to the distributive function of the price system.

Introducing the Price of Natural Capital by Changing the Distribution of Wealth

When our "extraterrestrial" wonders what value society should attach to the environment, he finds an extensive specialist literature by environmental economists: long discussions as to the merits of taxation or the allowances market, sophisticated calculations for determining environmental "shadow prices," questions about the discount rate and taking account of uncertainty, and so on. But when he lands on Earth and turns to practice, he discovers that one unanswered question dominates all the others. This question does not concern the state of the planet in twenty or a hundred years' time, but at the present moment. How can one find a new income distribution making the introduction of environmental value into the economy desirable here and now? How do people agree to change their income distribution so that the economy can take account of the value of natural capital? Once again, it is the problem of the shepherd and the owner.

Let us return to the prototype of the carbon price, which will undoubtedly play a key role in getting the ecological transition under way. Like the loss of biodiversity, climate change is an alteration of a global natural regulation system constituting a "common good." It is therefore out of the question to rely on an extension of the rules of private law to ensure its protection, as the economist Ronald Coase recommended in a well known paper:[3] the ownership of a coal mine or an oil well does not confer any additional right to emit CO_2, and it is impossible to price the damage by entering into a contract with people living in the proximity of the oil well or coal mine. Agreeing on the right level for this price and on its impact on income distribution is thus a political choice that will determine its social acceptability. Whether at the international or national level, the problem is to find the right tradeoffs in terms of distributional effects.

Chapter 13 showed how the distributional issue lay at the heart of international climate negotiations. A similar problem is encountered in establishing a carbon price within a country. In Europe, the classification of countries developed by Eurostat according to their levels of environmental taxation gives the highest rankings to the Scandinavian countries and Finland. These are also countries known for their art of social negotiation and that have resisted the vertiginous growth of inequality generated by the exuberance of financial capitalism. Is this a coincidence? Probably not. Detailed study of the Swedish carbon tax shows, for example, that the introduction of a high carbon tax compared to what has been done elsewhere in the world is based on a judicious and consensual balancing of distributional effects directed at lowering the charges weighing on companies on the one hand and targeting the poorest households with regard to social redistribution on the other.

In contrast, France has twice failed to introduce a carbon price into its economy: in 2001, on the initiative of Lionel Jospin, and again in 2009 on that of Nicolas Sarkozy. In both cases, legal action was to blame because the Constitutional Council accepted the appeals for its annulment brought by the opponents of carbon pricing. The basic reason was that on each occasion, politicians did not manage to achieve a consensus or explicit tradeoff on how to reinject, in an economically efficient and socially acceptable way, the proceeds from the new tax into the economy. A (fragile) consensus on the issue was reached in 2014, when President Hollande's government introduced such pricing on the basis of recommendations by a committee composed of economists and representatives of civil society.[4]

Pathways of the Circular Economy

One of the main virtues of a carbon price is that it facilitates the penetration of flow energies, a key factor in the transition toward a circular economy. The first industrial ecological pilot projects, such as Kalundborg eco-industrial park, opened in the 1970s in Denmark, revealed the importance of efficiency reservoirs formed by the recycling and re-use of industrial waste (and other energy sources). McDonough and Braungart's book *Cradle to Cradle: Remaking the Way We Make Things* introduced the concept of the circular economy to the general public.[5] The work of the Wuppertal Institute has shown its many possible applications, with the reliance on raw materials eventually disappearing as a result of widespread efficiency gains in their use, the recycling of end-of-life products, and the utilization of reproducible energy sources.

In *Factor Five* (2011), Ernst von Weizsäcker examines the economic reasons preventing the takeoff of the circular economy.

He points to the secular decline in the relative price of raw materials compared to the cost of labor and capital. This decrease stems from productivity gains in their production and use. To counter this effect, he proposes introducing a tax indexed to these productivity gains.[6] Such a tax would neutralize the transmission of efficiency gains to the price of raw materials, which would cease falling relative to the cost of labor and capital. Over time, it would become increasingly advantageous to use them sparingly and to recycle them.

The idea of taxing productivity gains is attractive to some environmentalists but unacceptable for economists. If the state automatically taxes productivity gains, it distorts one of the most powerful mechanisms for the spread of innovations in the economy as a result of lower prices. But the ecological transition needs more innovation. So it is best not to venture down this path. On the other hand, three types of incentives can greatly contribute to steering society toward a circular economy.

1) The first lever consists of incorporating the total cost of environmental damage into prices through taxation or cap-and-trade systems. In this regard, the development of pricing mechanisms for damage to biodiversity, along with the generalization of the carbon price, is the most decisive long-term incentive. One form of damage to ecosystems is the final disposal of waste in landfills, the pricing of which through taxation would greatly facilitate the creation of a circular economy. In chapter 11, we saw how fisheries, which were formerly destroying fish stocks, now operate in a circular economy system in which the price of fishing quotas provides an incentive to replenish the resource. By giving a value to the water purification service through the afforestation of the upstream catchment areas of the Hudson River, the City of New York also directly participates in this circular system. In relation to the general functioning of the economy, these experiments are still few and far between and have little impact on overall trends.

Their general application, as the right methods are developed for pricing the use of services provided by the diversity of ecosystems, is likely to profoundly change the situation. Biodiversity is indeed the lynchpin, prior to mechanisms for the reproduction of natural resources.

In this respect the example of photovoltaic panels used in the solar industry is emblematic. From the standpoint of energy, carbon pricing is the most powerful tool for effectively orienting investment choices. The price of carbon enhances the economic viability of this sector, as technical progress at the same time improves the energy efficiency of new-generation cells. Of course, such generalized pricing will negatively impact the cost of panels, insofar as their production and transportation uses fossil energy. If the carbon price is set at the right level, development choices will be made by reducing the impacts of the energy system on the equilibrium of the climate at the lowest cost.

But would carbon pricing be enough to orient the industry correctly? Like most electronic activities, the production and use of solar panels generate various types of environmental damage that interfere with the proper functioning of natural regulatory systems. The use of rare earths destroys biodiversity where they are mined; the production process requires the use of large quantities of very pure water, the discharge of which is not easy to treat; and the recycling of panels at their end of life gives rise to numerous technical problems. The right lever to encourage the sector's producers to take account of all these externalities affecting the balance of ecosystems is to introduce them into production costs through environmental pricing.

This example perfectly illustrates our point. For the photovoltaic industry, the real issue with regard to the circular economy is not to delay the prospective depletion of rare earths, but rather to integrate the whole production cycle into a virtuous circle that respects the various natural regulatory systems. To achieve this, the relevant economic incentives involve putting a price on any damage to these systems so as to integrate them into production costs. In the case of the photovoltaic industry, these incentives should steer R&D toward the perfection of processes with low environmental impact. In this regard, major qualitative changes could occur with the spread of bio-sourced solar cells.

2) Developing the circular economy is also a matter of switching from a product economy to a functionality economy. Under the necessity of economic constraints and competition, this shift is almost wholly implemented through trading between companies: all professional industrial equipment manufacturers are well aware that markets are now won on the basis of a comprehensive package—equipment and services—provided to the customer. With regard to households, the transition is being held back by decades of product marketing development and of powerful interests favoring the planned obsolescence of low-priced products, generally with poor health and environmental performance. In response to pressure from consumer organizations, increasing constraints are helping purchasers become better informed about the energy efficiency and water consumption of durable goods. This information signal is a first step toward the introduction of a price signal. When will pricing by usage become standard practice? In addition to its favorable environmental impact, such pricing would have a beneficial social impact because it is households with low purchasing power that suffer the most on account of the sometimes exorbitant cost of the poor performance of "lowest price" products.

3) Eco-design is the third lever for building up the circular economy. It involves incorporating, when designing products, the largest number of parameters for optimizing their use throughout the life cycle: manufacturing, operation, and disposal. A good economic incentive for optimizing usage and saving resources is to develop new forms of parafiscal charges allowing the price of the product to reflect not only its design and manufacturing costs, but also the outlay needed at the end of its life for storing in safe

conditions any waste that for technical reasons cannot be recycled. Such pricing gives economic meaning to the legal concept of "extended producer responsibility."

The Redeployment of Public Investment

Some authors, like Tim Jackson in *Prosperity without Growth*, consider the question of public investment in green technologies from a Keynesian standpoint and advocate a "green recovery."[7] When the authorities decide, as they did in 2009, to cushion the impact of a recession by increasing public investment, it is clearly preferable that the money involved is directed to projects that contribute to the ecological transition. Figures drawn up in 2009 by the bank HSBC suggest, however, that the majority of these investments (with the exception of Korea) went to the construction of sidewalks and highways, rather than protecting the environment. To suppose that the ecological transition will be set in motion through a sudden increase in public investment is thus a misdiagnosis.

On the one hand the time scales are not comparable. Keynesian reflation aims to counter short-term recessionary sequences, even if it means paying the unemployed to dig holes in the road and then fill them in again! The ecological transition is a long process of reorganizing resources resulting from the inclusion of the value of natural capital in the economy.

On the other hand, because of budgetary constraints it is unrealistic to count on additional public investment effort to get the ecological transition under way. Taking the environment into account has little chance of spontaneously leading to additional public investment, accumulating from one period to the next, for getting the economy back on track. We may believe or hope that this will occur, but building a policy on such an assumption would, at the very least, be rash. The state should instead support the strategic impact of the introduction of environmental pricing by redirecting its investments with regard to R&D and infrastructure.

Encouraging innovation to speed up the ecological transition gives rise to the so-called double externality problem, well described by Nordhaus (2013): "From an economic point of view, fundamental inventions have the same basic characteristics as global warming. . . . The only difference is that the externalities of innovations are largely beneficial while those of warming are largely harmful."[8] It logically follows that the public authorities need to step in and remedy this new market failure.

This issue primarily concerns the orientation of state-funded basic research. There can be no doubt about the role of such research in today's economy. Taking account of the ecological transition leads to a reconsideration of its priorities, so as to enhance our shared understanding of the fundamental mechanisms underlying the reproduction of natural capital. Such a reorientation has been initiated in favor of the various disciplines contributing to climate science, though it is still at an early stage with regard to the links between the life sciences and the reproduction of ecosystems. In the field of energy, for historical and military reasons, an altogether excessive proportion of basic public research is channeled into nuclear energy to the detriment of other low-carbon energies and energy efficiency.[9]

Between basic research and commercial exploitation lies an area that economists call the "Valley of Death," where the risk is too high to attract even the most adventurous entrepreneurs. Public support is therefore required to produce sufficient innovation. One common form of support consists of tax credits to encourage expenditure on research and development. Should these actions be specifically designated, for example by creating a "green tax credit" reserved for programs dedicated to the ecological transition? This green designation logic is easy to justify with the public and is well liked by the environmental movement. It is not optimal, however, with regard to public support for private R&D,

which would be much better guided by a price system credibly revealing the value of environmental goods over time than by administrative criteria specifying what is and is not green.

Another major area of government intervention concerns the choice of infrastructure, for which the judgment of the public authority is still decisive, even when arrangements involving private capital funding are used. Infrastructure determines far in advance how territories are organized and how resources may be allocated to them. When a city is built on the fragmented model of Los Angeles or Atlanta, there is no point trying to install a public transport network along the lines of Paris or New York. With the filling of Three Gorges Dam reservoir in China, the loss of land upstream is irreversible and the ecosystems downstream will be reacting for hundreds of years. What should be done to make infrastructure choices catalyze the ecological transition?

Economists' standard recommendation is straightforward: simply improve the economic calculations used for public investment choices. Unlike private investors, governments are not subject to criteria of profitability calculated from the market. They can therefore choose whatever theoretical ("shadow") values they want for pricing environmental externalities. To seriously address climate risk, one simply has to use a high carbon price and a low discount rate in the calculations to give sufficient weight to the future. In setting these new parameters, one obtains various options between existing infrastructures, and that's it!

Even if one makes the bold assumption that economic calculation plays a decisive role in public investment choices, the answer provided is not fully consonant with the question: the ecological transition requires not so much a different order of priority among existing infrastructures as an overall reconfiguration of their architecture. If one wants to build tomorrow's interactive energy system, one has to redesign each of the current system's basic links. Just as the closed network of seven million Minitel terminals produced by French telecoms could not be integrated into the Internet, it is impossible to develop a decentralized energy network while keeping a centralized power supply. At some point, a choice must be made between flow energies and nuclear energy, and between a territorial and a centralized approach.

As its value is progressively integrated into society, biodiversity will also become a powerful driver for the reconfiguration of infrastructure. It seems to have been forgotten that mangrove swamps, the afforestation of river catchment areas, and the diversity of life offer much more effective protection against major risks—such as tidal surges, the deterioration of water quality, the spread of epidemics, and so on—than concrete or antibiotics alone. Future governments will have to relearn how to use living organisms in infrastructure networks. This shift will entail depending less on economic calculation parameters than the ability to innovate in the context of decentralized territorial planning inspired by the forms of polycentric governance theorized by Ostrom.

Governance for Incorporating Long-Term Decisions

If we believe that democracy is the least bad system for representing collective preferences, it is our elected officials who have to shoulder this burden. They must contend with conflicting goals in the framework of electoral cycles that are not conducive to incorporating the long term into decision making. There is a serious risk of always postponing ambitious action to protect the environment or implementing stop-go policies in accordance with electoral cycles.

The problem is exacerbated whenever there is a question of coordinating action at the international level. To resolve disputes, some argue for the need to set up global political governance,[10] for example in the form of a world environmental organization with decision-making powers similar to those of the WTO. In view of the time required to come to a planet-wide agreement on such a body, we feel it's not something to get overly excited about!

Another proposed solution would be to have a planner under the supervision of experts mobilized to define the optimal solution, who would ensure that politicians carry out their work effectively and move in the direction of the predefined optimality. In practice, blindly relying on experts is almost certain to lead to social, economic, and environmental disaster. Structural adjustment policies conducted on the basis of best expertise have often shown their limitations in developing countries. An alternative (or complement) to experts is to introduce consultation processes with stakeholders, thus combining participatory democracy and representative democracy.

The right combination of experts, the market, planning, and stakeholders is hard to find. The experience of climate policies gives some indication of how it is desirable to bring these four types of actors together so as to reconcile democracy with decision making that incorporates the long view.

Rather than "experts," it would be better to speak of the state of scientific knowledge and the need to pass it on in a form accessible to policy makers and their stakeholders. The work of the Intergovernmental Panel on Climate Change (IPCC) led directly to awareness of climate change and the beginning of its integration into public policy. In a national context, the Committee on Climate Change in the United Kingdom plays a similar role with regard to the government and Parliament, thereby ensuring continuity of climate policy.

Such continuity has been sorely lacking in the management of the European CO_2 cap-and-trade system. The precipitate launch of the scheme between 2005 and 2007 showed that rapid progress could be made on the basis of decentralized bargaining in certain contexts. The rest of the story is less bright, with a series of shocks and failures that cost the instrument its visibility in the medium term. The system needs to be protected from interference and various short-term episodes by strong regulation, exercised by an independent authority receiving a clear mandate from the political authority. This type of architecture is similar to that of a central bank.

The implementation of climate (and energy) policies at the regional level is inextricably linked to new forms of urban and regional planning. Their structure cannot be derived from centralized architectures, which are totally incompatible with any consideration of the environment at the local level. Instead it should be inspired by the work of Ostrom, showing that the protection of the common good is never so well assured as when there is an organization that makes such protection wanted by all local actors. This process involves finding adequate forms of expression for these local stakeholders.

Consultation with stakeholders, particularly environmental organizations, has become required practice in developed countries when addressing environmental issues. In France, despite its tradition of centralization and the influence of the senior branches of the civil service, this type of consultation has become the norm with the Grenelle environment forum introduced by Jean-Louis Borloo. This major advance in terms of governance does not exclude raising some searching questions with regard to climate change. The risks associated with global warming primarily concern future generations and populations distant from France. The main stakeholder that should be consulted as to its preferences, its degree of acceptance or aversion to this risk, and its propensity for altruism is in point of fact absent, namely the next generation.

In the real world there are, however, many contenders prepared to invite themselves to the table and represent the interests of this coming generation. Attempts are often made to compensate for the absence of the next generation by consulting a plethora of stakeholders. Hence the risk, inherent in multi-actor and multilevel governance, of delaying decision making or, worse, constantly deferring any decision in the name of necessary democratic debate. Governance that takes the long-term view is by no means synonymous with governance that prolongs decision-making processes.

19

European Strategy

Jump Out of the Warm Water!

JUST AS THE CLIMATE issue has its climate change skeptics, so too Europe is faced with the rise of Euroskepticism. The problem seems to have worsened since the Great Recession of 2008 and its knock-on effects. Some believed that the economic and financial crisis triggered by the bursting of the speculative bubble on Wall Street would lead to a rebalancing between the "neo-liberal financialized" capitalism promoted by the United States and the more social and regulated capitalism favored by Europe. As a whole, Europe certainly fared better at the very beginning of the recessionary shock of 2008, limiting the decline in production through the triggering of "automatic stabilizers,"[1] with only some peripheral countries such as Greece and Ireland plummeting abruptly. It then became clear that the single currency was not the shield it was hoped to be. In having to deal with the interminable euro crisis, Europe has become bogged down in a recessionary mire, whereas the investment banks on Wall Street have recovered their self-confidence, with an economy reinvigorated after liquidating some of its excessive debt.

The same doubts are apparent regarding climate and environmental policies. Since the Copenhagen summit, Europe seems paralyzed in terms of climate policy. It has not been able to agree on a new set of objectives for 2030 and is letting the European CO_2 emissions trading scheme drift. The complexity of the Climate and Energy Package adopted in 2008 and regulatory instability discourages investment in the energy transition. Doubt is becoming more pronounced in the European parliament and the European Council, where climatic and environmental ambitions vie with concerns about competitiveness and employment. Doubt is also increasingly being expressed by voters, who in the elections to the European Parliament in May 2014 returned an unprecedented number of MEPs hesitant about, or downright hostile to, European integration.

The valorization of green capital may be the best antidote to such doubts and may constitute an engine facilitating the restructuring of the European economy. For this to happen, it is essential to stop playing a waiting game with regard to climate change and not succumb to the "frog in warm water syndrome." The integration of environmental values leads to a new, more efficient economy, more capable of attracting people's support. To enter this new phase of the European project, the European Union should reaffirm its climate ambitions and refocus its energy transition policy on carbon pricing. It should also promote an industrial competitiveness strategy based on energy efficiency and the building of new industrial sectors, and integrate its agricultural policy into a new ecological management of territories.

The Risk of Losing on Both Fronts

The "frog in warm water syndrome" consists of failing to draw all the consequences of the strategic choices one has made and seeking to win on all fronts. At the macroeconomic level, Europe has paid dearly with the euro. A single currency cannot function properly without an integrated economic area with high mobility of factors of production, including labor, in order to equalize costs, and close coordination of fiscal policies. By wanting to benefit from the advantages of the single currency without accepting all its implications, the euro area underwent a near fatal crisis between 2008 and 2014, with very high adjustment costs in terms of unemployment.

Regarding the energy transition, Europe is developing this same syndrome on a grand scale and failing to take full advantage of its still considerable strengths. Among industrialized countries, the European Union has the best energy efficiency ratio (the amount of energy used as a proportion of GDP). With the average carbon content of the energy it uses lower than in other countries and having fallen sharply since 1990, it emits only seven tons of CO_2 per capita, compared to seventeen for the United States and ten for Japan and Korea. Its known fossil fuel deposits are already considerably or completely depleted, putting it in a very different situation from the United States, with its unconventional hydrocarbon revolution.

In both the European Parliament and the European Council, an increasingly frequent refrain is heard whenever the issue of the transition to a low-carbon economy is raised: pursuing a low-carbon strategy would expose Europe to growing risks for the competitiveness of its industry. The increased competitiveness of American industry, which has benefited from a sharp fall in the price of gas, is then singled out as a prelude to the clinching argument: the risk of increasing relocation to emerging countries. Expressed more bluntly, China, by not encumbering itself with climate considerations, will be the big beneficiary of EU policy, which, though probably morally right, is suicidal for EU's industrial future.

The argument merits attention. As we have seen, the introduction of prices to internalize environmental damage is not intended to place additional burdens on the economy, but to generate transfers between economic agents. In industry, these transfers will be made from energy-intensive mature industries, in which it is difficult to reduce emissions, to the advantage of industries developing low-carbon technologies. The risk of the former relocating to areas where CO_2 emissions are free, rather than reorganizing their production system, is real enough and needs to be managed. But for all that, should carbon pricing really be sacrificed to satisfy industry lobbies concerned about reassuring their employees?

In proceeding in this manner, Europe risks losing both ways. Its warm water policy will not save those parts of the traditional industry that have no other choice but to opt for energy efficiency and moving upmarket if they are to remain internationally competitive. At the same time there is the risk of being overtaken by Asian and American competition in the emerging green economy niches. While the European Parliament has been discussing the impact of the price of a ton of CO_2 on the competitiveness of the European steel and cement industries, China has been taking a leading position in the photovoltaic cell industry. Renewable energy, and green industry more generally, has in fact been placed at the forefront of China's economic priorities in its twelfth five-year plan.

Revising the Climate and Energy Package to Boost Investment

For years, European energy policy has essentially been to liberalize national markets for electricity and gas while seeking to integrate them through interconnections between countries. In parallel, Europe has introduced rules designed to harmonize taxation on energy products by setting minimum thresholds for excise duties in order to limit the possibilities of fiscal dumping.[2] It finally launched the CO_2 cap-and-trade scheme in 2005, the first market to operate with fully harmonized European rules. Most citizens know very little about any of these technocratic constructions, which have not been explained to them and on which they did not have a say.

The adoption of the climate and energy package in 2008 was the first attempt to link these different elements within a medium-term proactive perspective. With what is commonly known as the "3 times 20" target, the idea is to guide energy choices in accordance with three shared objectives for 2020: a 20 percent reduction in greenhouse gas emissions compared to 1990; a 20 percent increase in the share of renewables in the energy mix; and a 20 percent gain in energy efficiency. The package of measures is broken down by country and sector in a series of texts amounting to over a thousand pages in all,[3] the complexity of which ends up losing even the experts themselves. The result is a tangle of objectives and instruments that undermines the effectiveness of the system and disrupts the functioning of the electricity market, as Boris Solier clearly shows.[4] The most obvious consequence is that instead of setting medium-term way markers to guide economic actors, this system confuses the issue and discourages investment in the energy transition. It is therefore essential to simplify the climate and energy package and make it more comprehensible, by specifying in particular what the responsibility of Europe is and what should be left to the initiative of Member States and regions.

One of the most important issues to resolve is the harmonization between climate objectives and those of traditional energy policy. The current system leads to the undesirable superimposition of instruments. On the one hand, the goal of reducing emissions by 20 percent by 2020 has to be attained at least cost through the carbon market, presented as the epicenter of the European climate strategy. On the other, the legislator has introduced binding penetration rate targets for renewable energy sources by country and by energy category for 2020. This second objective is not supported by a shared economic instrument, but has given rise to the development of national support schemes for renewable energy through feed-in tariffs, which distribute high rents to producers, financed either by taxes or by consumers as a whole.

This lack of coordination often leads to absurd outcomes. Germany is fast developing its production of electricity from renewables, thereby weakening the price of CO_2 allowances in the market. Then suddenly, in the absence of a sufficiently high carbon price, it closes or suspends its gas-fired plants and invests in new coal-fired plants! To avoid this lack of coherence, the United Kingdom is superimposing on the Community system a national tax on CO_2 emissions in its electricity sector. Though consistent with its own interest, this further destabilizes the EU emissions trading scheme. As shown by the work carried out under the direction of Raphaël Trotignon,[5] the superimposition of these instruments, along with the lack of real political support, is the underlying reason for the shortcomings of the CO_2 allowances system.

To initiate its energy transition under good conditions, Europe must choose between two strategies. If it relies on the foresight of governments and their experts to decide the right energy mix in advance—the logic of the Renewable Energy Directive—it should favor an approach based on planning. In such an approach, the European emissions trading system is no longer the main instrument and can be abandoned without regret. This would be a possible and not irrelevant option, but it is not one we recommend. If, on the other hand, a transition by means of economic incentives is preferred, carbon pricing must become the central lever stimulating the investment required to initiate the energy transition. Public support for renewables must then be reconfigured within the framework of a supply-side policy based on strengthening industrial competitiveness.

Carbon Price Markers

In view of the problems that have repeatedly arisen in the European carbon market, some economists advocate changing the method by pricing CO_2 emissions by means of a tax, in accordance with the procedure presented in chapter 12. The advantages of such a policy change may be debated at length. In practice, it has no chance of being implemented because such a decision would involve a politically unobtainable, unanimous decision by the twenty-eight Member States. Within the existing framework, there are two possible channels for reinvigorating European carbon pricing: restoring the credibility of the carbon market for industrial emissions covered by the allowances system, and introducing national taxes in other economic sectors.

The turnaround of the emissions trading scheme first requires the involvement of heads of state to give political credibility to the system by adopting ambitious emissions reduction targets for 2030 (probably a 40 percent reduction). Such an agreement faces opposition from some Member States because of the widespread fear of possible weakening competitiveness. As we have seen, it might be easier if progress were to be made in international climate negotiations. The other condition for the recovery of the ETS is the establishment of new governance based on an independent authority, which, like a central bank, would ensure the long-term stability of the framework while providing the flexibility required for adapting to short-term contingencies.[6]

Extending the carbon price to the rest of the economy has in each case depended on the country concerned. For reasons of political balance between different Directorates-General, the Commission decided not to extend the CO_2 emissions trading scheme to transport and construction, and is proposing to use taxation to price emissions linked to their various uses. This hybridization between a market tool and a tax unnecessarily complicates carbon pricing and reduces economic efficiency by multiplying carbon price signals within Europe. Considerations of efficiency will therefore point to the need to expand the sectoral field covered by the market, once this has been rectified.

To lend credibility to the system in the long term, it is also important to be transparent about the distributional effects induced by carbon pricing and to use them to direct resources according to the general objectives pursued. As shown by the example of the shepherd and his employer, the right compromise regarding redistribution has to be found.

What sorts of figures are we talking about? If it costs €20 to emit a ton of CO_2 into the atmosphere, Europe would currently value a stable climate at €100 billion. A broad consensus of economists argues that the price required to guard against the risks of overly rapid warming should quickly approach €100, a figure that would initially bring the carbon rent up to €500 billion. Once the carbon price catalyst is taken seriously, it is clear that it involves appreciable amounts at the macroeconomic level. It is for this reason that the introduction of ambitious pricing of climate into the European economy goes hand in hand with a strategy of redistribution of environmental rent at three levels: geographic, economic, and social.

Distributional Impacts and Social Involvement

Firstly, carbon pricing can become an instrument facilitating economic transfers among countries. Trotignon and Delbosc estimate that transfers made to Eastern Europe in the early years of the functioning of the carbon market amounted to €500 million through the sale of CO_2 allowances distributed free of charge.[7] With the move to auctions, it was decided to return most of the financial product to Member States, thereby limiting the possibility of using this money to make lump sum transfers between countries. The climate emergency has nothing like the impact of financial crises, which lead to much greater sums being hurriedly transferred between countries.

The most visible distributional effects of existing carbon pricing are those relating to industry. As we have already mentioned, these distributional effects have up until now favored companies subject to quotas that have benefitted from the free allocation of their emission allowances. In the electricity sector, in which the art of capturing rent is the lifeblood of competition, this has resulted in windfall profits. In the industrial sector, the system has subsidized basic intermediate industries such as steel and cement to a very considerable extent. The Arcelor-Mittal group, for example, received a net transfer of several hundred million euros during the first eight years of the market. If we want to hasten the energy transition, it will be necessary to rebalance these sectoral transfers and in particular to subject them to effective reconversion toward low-carbon processes and economic sectors.

The extension of carbon pricing to activities currently not covered by the emissions trading system will affect households more directly. In the medium term, the rise in prices will be at least partially offset by efficiency gains in usage and switching to flow energy. But in the short term, what should be done with regard to the sixty-five million European households (13 percent of the total) considered by the Commission to be in a situation of "fuel poverty"? And most importantly, how can the 20 to 30 percent of households that are not currently "precarious," but for whom energy costs already weigh heavily on their monthly outlay, be prevented from becoming so? Europe has so far carefully avoided this issue by focusing on industrial emissions and not touching emissions from transport and housing that have the most direct impact on household budgets. Consequently, the extension of carbon pricing to these sectors is making no headway. The majority of Member States are opposed to such an extension because it would give rise to further social problems in addition to all those already induced by the crisis.

To escape this inertia, it is essential to rigorously assess the social impacts of climate pricing and to introduce accompanying measures. Authors such as Jean-Marie Chevalier, Michel Derdevet, and Patrice Geoffron advocate a "European energy shield" to cushion the "shock of the transition energy on the most vulnerable households."[8] Such a shield would involve transferring financial resources to lower income households in the short term and targeting these households with ways to improve efficiency and facilitate energy substitution in the medium term. The feasibility of such a shield is, however, almost zero due to the current institutional organization of Europe, which makes social issues a reserved area at the sole discretion of Member States.

But is this organization set in stone once and for all? At major points in its history, the European Community has shown that it can modify its institutional organization and achieve greater unity in response to common priorities. Such a shift occurred, for example, in the construction of the common agricultural policy in the 1960s, leading to the sharing of funding that still accounts for more than half the EU budget. The introduction of the euro and the management of its subsequent crisis also led to considerable pooling of resources. If the climate is a true priority for Europe, it must also entail a pooling of means with a view to procuring ambitious carbon pricing. For its climate mitigation program, Europe needs to have a strong financial arm if it is to fairly manage the redistribution engendered by carbon pricing and strengthen its industrial sectors.

The Energy Transition and Competitiveness: Stimulate Supply or Protect the Borders?

The debate around the linkage between the energy transition and competitiveness often tends to focus on the possibility of neutralizing the impact of the carbon price by means of a levy mechanism at Europe's borders that would put traditional industries on a par with their foreign competitors. But while the well-worn banner of carbon pricing at borders is being brandished, Chinese industry has been making a historic breakthrough in solar photovoltaic energy, an industry that would not have been in the least hampered by such a measure. Making the energy transition a factor for industrial revival is not so much about border measures as it is about implementing a supply-side strategy to boost investment and innovation.

The example of using "feed-in tariffs" to facilitate the development of photovoltaics is a reminder of the damage that can result from a lack of strategic vision on the subject. Energy producers are fond of such tariffs, as they guarantee in advance a minimum price for renewable electricity or biogas, in other words a rent incorporated into the guaranteed price. This type of mechanism recalls the heyday of the common agricultural policy, when Europe guaranteed the producer an intervention price for wheat or butter, regardless of market conditions. In Europe, feed-in tariffs for renewable energy have been introduced by many Member States since 2005, under the impetus of a coalition of environmental NGOs and private interests. The sector in which these tariffs have been most generous is solar photovoltaics, in which a lucrative rent funded by uncoordinated national policies is distributed to producers. The only problem is that it is Chinese manufacturers, much better organized and coordinated than their European counterparts, who have run off with all the prizes.

Standard solar panels mounted on rooftops or in arrays on the ground represent between a quarter and a fifth of the added value of the solar industry. They are all fabricated in highly capitalistic production systems in which size and the ability to quickly deploy electronic technologies are the conditions for success. With a tenfold increase in installed capacity in the world, the photovoltaic market took off between 2006 and 2012, mainly due to the incentive of European feed-in tariffs. Nonexistent in 2003, Chinese manufacturers held more than three-quarters of the market in 2012, having induced the bankruptcy or bought up the majority of European producers. The cost of labor had nothing to do with this success. The crucial lever was the massive provision of credit to Chinese manufacturers by local state-owned banks on unbeatable terms: true financial dumping!

Almost a year later than the United States, in March 2013 the European Union decided to set up a registration procedure for imported products in the sector liable to duties under anti-dumping measures. But it is clear that such border protection only addresses the effects of the lack of foresight by Europe, which deployed a powerful incentive to develop the use of new energy without supplementing it with the required means to deploy a competitive offering. Europe should not be naive when it comes to applying customs duties. But there are far more effective ways to make the energy transition a lever for innovating and reinvesting in industry.

In the industrial sector, part of the carbon rent should be used to develop new green industries and redeploy traditional industries toward low-carbon production methods. The European NER-300 program is the embryo of such a policy. It enables some of the proceeds from CO_2 allowances auctions to be recycled in innovative projects in renewable energy or carbon capture and storage. With such a recycling of carbon rent, manufacturers, including those in traditional sectors, have a direct stake in the increase in the price of carbon, as it gives them more resources with which to innovate and invest.

Let us now turn to an example that is emblematic in France: the Lorraine steel industry. The competitiveness of the blast furnaces at the historical sites of Lorraine has been compromised by the additional transport costs of ore and coal compared to port facilities. If the industry limits itself to standard production techniques, no manufacturer can seriously hope to stay in business. The question arises differently if the site were to be transformed into an industrial pilot for testing new production techniques to enable emissions to be drastically reduced through carbon capture and storage. This is precisely what the European Ulcos program is aiming to do. Economically, ramping up the pilot scheme becomes profitable if the price of carbon is high enough for it to become worthwhile to capture CO_2 as it emerges from the stacks and then store it. Contrary to what some people would have us believe, the competitiveness of the Lorraine steel industry is therefore in no way burdened by the carbon price. Indeed the opposite is the case: its future is dependent on more ambitious carbon pricing that will enable steel to be produced in one of France's oldest industrial areas in the future, with virtually no CO_2 being emitted.

Energy Efficiency, the Engine of Competitiveness

We cannot emphasize it enough: the first condition for the successful replacement of fossil fuels by flow energies and biomass is a drastic acceleration of efficiency gains in the use of energy. This acceleration requires numerous technical and organizational innovations, many of which will be built in a decentralized way at the local level. Energy efficiency also makes a major contribution to competitiveness, reducing or eliminating the uncertainties faced by industry with regard to its energy supply.

The promotion of energy efficiency goes hand in hand with providing information to citizens and consumers. It should also exert a stronger influence on Europe's policies around energy networks and security of supply. The grid interconnectedness policy is certainly useful, but it is no longer the main issue for the future. The main issue is the multipolar and interactive network revolution, for which Europe urgently needs to lay the groundwork, both from an institutional perspective and with regard to technical standards. Energy efficiency (and the flexibility it requires in the short term) should, moreover, become the principal lever to strengthen the security of supplies.

Europe lacks the economic tools to provide powerful incentives for energy efficiency. In this regard, it should mobilize two instruments still embryonic on the European scale: third-party investment and energy saving certificates.

Third-party investment involves using the future value of economies generated by an efficiency program to fund that program. Many pilot projects have been conducted by the International Bank for Reconstruction and Development (IBRD) in Eastern Europe and by KFW in Germany, two public development banks. Third-party investment is not growing in the financial system because it is too risky: as well as the operational risk—the effective implementation of the program—there is the further risk around the future price of energy, which determines the value of the economies made. And as energy economists know better than anyone, that price cannot be predicted. To deploy large-scale third-party investment in Europe, there is a well-defined route: the European Investment Bank could initiate and coordinate third-party operations in partnership with national public financial institutions. Major involvement on the part of these institutions would be likely to reduce risks and break down many barriers within the financial system.

Though still in its early stages, the instrument of energy saving certificates has been developed in rather different ways in more than a dozen countries in Europe, including France. The principle is very simple. Instead of using a strictly regulatory approach requiring each individual actor to make a specified amount of energy economies, it sets an overall savings goal that is then apportioned among energy distributors. The distributors can then mutually trade certificates, each of which attests to a specific amount of energy savings. If the trading of certificates is organized transparently and with sufficient liquidity, it gives rise to a certificate price that may encourage actors outside the system to enter the market and thereby improve their own energy efficiency. Developed in an uncoordinated way at the national level, this type of instrument has little chance of taking off. Here again, Europe should take the leap by creating the conditions for the emergence of a true European market in which energy savings will in the future be valued in accordance with the opportunities arising in each territory.

Environmental Management of Territories

To give impetus to the ecological transition, Europe must abandon half-measures and introduce powerful incentives to rapidly shift its energy system toward a low-carbon model. These economic incentives acquire their full meaning within the framework of strengthened international coordination.

At the same time, local links must be strengthened by taking better account of the diversity of territories and through the emergence of decentralized initiatives. True for energy, it is even more so with regard to the other major control system of natural capital: biodiversity. Such taking account of territorial levels is complex in terms of governance and raises many issues in terms of equity between regions. But in Europe there is a large-scale sectoral policy whose redeployment could eventually become a major driver for the environmental management of territories: the Common Agricultural Policy (CAP).

Created originally as a highly effective system for supporting agricultural prices in a trade deficit context, the CAP has undergone ceaseless reform since the mid-1980s.[9] Because these reforms consist largely of eliminating rent seeking, which has historically benefitted certain producers (and many of their suppliers and service providers), the reform movement is not very popular in the world of agriculture. Up until now it has wholly lacked a mobilizing vision, simply seeking to reduce the subsidies to agriculture by decoupling them from the amount of produce marketed.

If they are skillfully introduced, economic incentives to invest in climate protection and biodiversity could significantly change the situation. Reorienting agricultural development in accordance with the climate objective would entail introducing a carbon price as an instrument to support agriculture. Similarly, advances in the assessment of losses or gains in terms of biodiversity should allow the sector to be managed more efficiently by gradually changing direct aid into payment for ecosystem services. A policy for promoting investment in biodiversity would thus be gradually established, which would be much more effective than current actions to limit access by means of natural reserves.

We are aware that setting up experiments in this area and then implementing them cannot be done overnight. But by gradually integrating the values of climate protection and biodiversity into agricultural policy, Europe can initiate a movement toward ecologically intensive agriculture models that the world will probably very much need in the future. In the long run, this is perhaps one of the most profitable investments in the ecological transition.

CONCLUSION

Green Capital, Green Capitalism?

IN THE INTRODUCTION WE raised the question of the color of growth. Now, as we come to the end of our journey, we return to the color of capital. The central thread of our argument has been the introduction of a new factor, alongside labor and productive capital, into the production function, namely natural capital. In the standard representation, economists view natural capital as a limited stock of resources whose scarcity may hamper growth in the long term. We have tried to improve this representation on the basis of a simple idea: less than the scarcity of raw materials, it is the deterioration of the great natural regulatory functions, such as the climate system through the greenhouse effect and biodiversity, that endangers continued expansion. To remedy this situation, it is essential to introduce into the economy a new price that measures the cost of the damage to these regulatory functions. This new value changes the relative prices of factors of production and disrupts the existing distribution of revenues. It triggers new investment in pollution reduction so as to rebuild natural capital. Investment in green capital thus becomes a springboard generating multiple innovations that reorient growth by ensuring its sustainability. From one period to the next, this affects the pace and composition of the accumulation of capital stock in the economy.

How is this new accumulation path positioned compared to past phases of capitalism? We reject outright the simplistic idea conveyed by some green growth activists that this reorientation of the accumulation process takes us spontaneously to a fairer, more altruistic, more inclusive society. The example of the shepherd and his employer is there to remind us that there are many ways of sharing environmental rent once the value of natural capital is introduced into the economy. Are we then unconsciously paving the way for increased exploitation of labor by a triumphant green capitalism? Is the search by economists for ways of pricing environmental damage, citing the superiority of decentralized regulation by price over the more or less tyrannical choices of a planner, little more than a cover for tomorrow's rampant capitalism?

Since 1945, the accumulation of capital has grown at an unprecedented pace globally. As Thomas Piketty points out in his *Capital in the Twenty-First Century*,[1] this process has been accompanied since 1980 by a twofold dynamic: a reduction of the income gap between high-income and low-income countries, and increased inequality in the United States and Europe to a degree not seen since the beginning of the last century. When the polarization of wealth becomes such that the wealthiest 1 percent account for over a fifth of the national income, as in the United States in the early 2000s, it is likely to have consequences that will transform the existing order. Such a situation occurred only once in the last century: in 1928. The shock that followed in the form of the Stock Market Crash of 1929 destroyed vast amounts of capital and permanently altered fiscal policies.

While Piketty skillfully recounts the rise of inequality linked to past capital accumulation, the rest of his story is somewhat demotivating. He refers to a shock measure that could change the situation: the establishment of a global capital tax, at a progressive rate, to bring an end to the phenomenon of the intergenerational transmission of wealth. The introduction of such a tax has about as much chance of happening as everyone waking up tomorrow morning and finding a single universal carbon price in place! Piketty recognizes this because he describes such a tax as purely utopian and warns that the society of the twenty-first century, with a decline in the rate of growth that will further reduce the return on capital, promises to resemble the one described by Balzac, a society in which inheritance increases inequality from one generation to another, and in which it is better to marry a rich heiress than pursue one's studies.

The shock resulting in a redeal of the cards could well be environmental. Society is only beginning to become aware of the costs engendered by the deterioration of natural regulatory systems. These costs are still small compared to those that future generations will pay, especially if one takes into account the risk of the collapse of certain regulatory functions that may occur at various "tipping points," in the same way that withdrawals by small savers can precipitate the collapse of the entire financial system in the event of bank runs. If the scenarios of the breakdown of the Amazonian ecosystem or the melting of the Greenland ice sheet were to materialize, who has the slightest idea of the costs involved?

The introduction of large-scale environmental pricing to gradually incorporate these costs into the functioning of our economy is a great opportunity to correct the many excesses secreted by financial capitalism since 1980. The mobilization of green capital can indeed result in a real transformation of the capitalist system for three basic reasons.

The first is that the ecological transition will change the composition of productive assets. The dynamic aspect is that new assets—technological, human, organizational—will gradually be formed due to the new scale of relative prices. As the value of natural capital becomes embedded in the economic system, it will be evident that some of the existing assets have become too expensive in the new scale of values because the cost of polluting hampers their performance. They will be downgraded and disappear from the stock of capital. This destruction of "dirty" productive capital will be a lever enabling the average rate of profit to be maintained and will redeal the cards pertaining to the inheritance of wealth. All major changes in capitalism have always gone hand in hand with this process of the destruction and reconstruction of capital, or "creative destruction" as Joseph Schumpeter termed it.[2]

Pricing the environment does not have the "punitive" aspect of a straightforward tax on capital, and this facilitates its acceptance by society. It is a matter of protecting a set of common goods, both global and local, needed by the rich as much as the poor. Pricing the environment and a tax on capital have one characteristic in common: they both concern the relations between generations. But the legacy in question is no longer like that of the young Rastignac (in Balzac's *La Comédie Humaine*), to be dissipated after he seduces his wealthy heiress. It is not a matter of punishing Bill Gates and Warren Buffet or their heirs, but of protecting a legacy we pass on to all future generations. This legacy increases the solidarity between generations and at the same time enables them, through the new growth opportunities it will create, to escape the weakening of economic expansion.

By construction, the introduction of natural capital in the economy is based on the principle of solidarity. Might this principle be accompanied by increasing inequality within the current generation? From a strictly theoretical standpoint, one could model such a phenomenon, given various rather unrealistic conditions. In reality, today's society is confronted by many obstacles that hinder the mobilization of green capital: combatting pollution is too expensive; it undermines competitiveness by weighing on production costs; it's a luxury that must be reserved for the rich. These clichés orchestrated by various lobbies continue to blind our policy makers. To counter them, we must find ways of redistributing environmental rent that stimulate the economy and mobilize the greatest number of people. The ecological transition can therefore become a powerful tool for social transformation and reducing inequalities. Similar movements have, moreover, been seen historically with the mobilization of society during periods of reconstruction after wars or disasters. An interesting finding from simulations of the introduction of the carbon price into the French economy is that it concomitantly brings about transfers from older generations to younger generations within existing society.[3] Because older people have on average become much richer in Western society than the young, this reduces inequality.

The succession of changes resulting from the mobilization of green capital leads to a society in which a new balance will be found between the market and government regulation, both globally and in relation to territorial management. In this model of society, markets continue to play the leading role in the allocation of resources. Thus it can hardly be described as "socialist." Let us say, rather, that it is a "post-capitalist" model with many possible variants, because the ecological transition, with the aim of changing the relationship of mankind to nature, is also likely to transform social relations.

Over and beyond our perspective as economists, there are many other reasons to reconsider how society needs to protect its green capital. Prices, one of most effective instruments, are only an extremely partial measure of society's values. Holding natural capital in greater respect is a requirement that squares with many other values: cultural, ethical, and aesthetic. With the extraordinary growth of technology, mankind has built an artificial world, using nature as a kind of freely available toolkit—as described by Descartes—without the least concern about the reproduction capacity of this natural system. We are now superimposing on this artificial world a virtual world that constantly increases its hold over our behavior and representations. These artificial constructions cloud our relationship with the natural world and give us dangerous illusions of power. Such illusions make us forget that mankind is but one link in this world or, in the terminology of Spinoza, Descartes's great antagonist, but one of the attributes of substance. It is for this reason that building a new relationship with nature also responds to a deep yearning for wisdom. Our contribution as economists has simply been to show under what conditions the mobilization of green capital can enable this goal to be reached at the least cost.

Notes

Introduction. The Color of Growth

1. The first examples of this primitive writing system known as rong-orongo were discovered on wooden tablets in 1864 by a missionary. The script has so far not been deciphered.

2. Jared Diamond, *Collapse: How Societies Choose to Fail or Succeed* (New York: Penguin Books, 2005).

3. TEEB, *The Economics of Ecosystems and Biodiversity: Mainstreaming the Economics of Nature: A Synthesis of the Approach, Conclusions and Recommendations of TEEB* (TEEB, 2010).

4. The book by Donella H. Meadows, Dennis L. Meadows, Jorgen Randers, and William W. Behrens, *The Limits to Growth* (New York: Universe Books, 1972), was unfortunately published in French under the title *Halte à la croissance* (Paris: Fayard, 1972). Republished in 2004 in the United States, it was followed by a new French edition under the title *Les Limites à la croissance* (Paris: Rue de l'Echiquier, 2012).

5. The term was coined by the French economist Jean Fourastié. It designates the thirty years of rapid growth following the end of World War II. The American economist Angus Maddison speaks of the "Golden Age" in reference to this period.

6. David Pearce, *Blueprint for a Green Economy* (London: Earthscan, 1989).

1. Growth: A Historical Accident?

1. The analyses are based on the "World3" model, the most comprehensive documentation of which can be found in Dennis L. Meadows et al., eds., *The Dynamics of Growth in a Finite World* (Cambridge, MA: Wright-Allen Press, 1974).

2. The Bretton Woods agreements provided fixed parities between Western currencies and the U.S. dollar, itself convertible into gold at a rate of $35 per ounce.

3. For an insightful and highly informative analysis of this crisis, see Michel Aglietta, *La Crise, les voies de sortie*, 2nd ed. (Paris: Michalon, 2010).

4. The term refers to real estate debt, which in the United States has been massively transferred from banks to the capital market by the technique known as "securitization" (conversion of bank debt into financial securities).

5. See, in particular, Angus Maddison, *The World Economy: A Millennial Perspective* (Paris: OECD, 2001), and Paul Bairoch and Maurice Lévy-Leboyer, *Disparities in Economic Development since the Industrial Revolution* (New York: MacMillan, 1981).

6. This philosophical system was founded by Auguste Comte in his *Cours de philosophie positive*. It considers that only established facts have universal value and lead to the codification of knowledge derived directly from observation and experience.

7. This point has been the subject of a bitter controversy between Bairoch and Maddison, who were nevertheless in agreement on the diagnosis that there was little to choose from between the two economic blocs in comparison with the situation in the mid-twentieth century.

8. The concept of peak oil was developed by the geologist M. King Hubbert from observation of the operations of American oil wells. The "Hubbert curve" shows that oil extraction becomes expensive and declines from the time at which half the reserves have been extracted. It is on the basis of this curve that some people regularly announce the end of the "oil adventure." For a well-documented presentation of these theses, see Adolphe Nicolas, *Énergies: Une pénurie au secours du climat?* (Paris: Belin, "Pour la science," 2011).

9. See Robert Gordon's meticulous analysis of the digital revolution "Does the New Economy Measure Up to the Great Inventions of the Past?," *Journal of Economic Perspectives*, 14 no. 4 (Fall 2000): 49–74, and his more recent paper "Is US Economic Growth Over? Faltering Innovation Confronts the Six Headwinds," *CEPR, Policy Insight*, 63 (September 2012): 1–13.

10. Gordon, "Is US Economic Growth Over?," 7.

2. The Spaceship Problem: An Optimal Population Size?

1. K. Boulding, *The Earth as a Spaceship* (Baltimore, MD: Washington State University Committee on Space, 1966). www.colorado.edu/econ/Kennneth.Boulding/spaceship-earth.html.

2. G. Botero [1588], *Delle Cause della Grandezza della Città*, partly reproduced in *The Population and Development Review* 11, no. 2 (1985): 335–340.

3. R. Cantillon, *Essay on the Nature of Trade in General* (London: Frank Cass and Co., 1952).

4. T. R. Malthus, *An Essay on the Principle of Population* (London: J. Johnson, 1798).

5. Jeremy Bentham, *An Introduction to the Principles of Morals and Legislation* (Oxford: Clarendon Press, 1789).

6. H. Sidgwick, *The Methods of Ethics* (London: MacMillan, 1874).

7. D. Parfit, *Reasons and Persons* (New York: Oxford University Press, 1984). Parfit argues that utilitarian views suffer from a risk of a "repugnant conclusion." In other words, for a population of a given size, there exists one of a larger size with a lower level of per capita welfare but a higher overall level of welfare.

8. "All other things being equal" (or "ceteris paribus") is a favorite expression among economists because it allows a given phenomenon to be isolated and understood by temporarily ignoring other interactions.

9. J. Broome, *Weighing Lives* (New York: Oxford University Press, 2004).

10. See P. A. Jouvet and G. Ponthière, "Survival, Reproduction and Congestion: The Spaceship Problem Re-Examined," *Journal of Bioeconomics* 3 (2011): 233–273.

3. Degrowth: Good Questions, Bad Answers

1. John Stuart Mill, *Principles of Political Economy* (London: John W. Parker, 1848), bk. 4, chap. 6.

2. The rebound effect, or the Khazzoom-Brookes postulate, refers to the paradox highlighted by W. S. Jevons, who in 1865 observed a sharp increase in coal consumption despite significant efficiency gains in its use. Efficiency gains do not, therefore, lead to a decline in consumption, but rather to an overall increase due to lower costs. See Harry D. Saunders, "The Khazzoom-Brookes Postulate and Neoclassical Growth," *The Energy Journal* 13 (1992): 131–148.

3. See, in particular, Michel Griffon, *Pour des agricultures écologiquement intensives* (Montrouge: L'Aube, 2011).

4. Tim Jackson, *Prosperity without Growth: Economics for a Finite Planet* (London: Earthscan, 2011).

5. As the Scottish economist Robert Giffen noted, they can in some cases respond "backwards": when the price of an essential commodity (the so-called Giffen good or inferior good) such as wheat or potatoes rises, its demand by low-income households may increase because the increase in price makes it impossible to use a second "superior" foodstuff, such as meat or fresh produce.

6. See the report by Benoît Bréville, "Obésité, un mal planétaire," *Le Monde Diplomatique*, September 2012.

7. Marion Nestle, "Does It Really Cost More to Buy Healthy Food?," www.foodpolitics.com, August 5, 2011.

8. Jackson, *Prosperity without Growth*, 175.

9. See the report "For Richer, For Poorer," *The Economist*, October 13, 2012.

4. Introducing the Environment Into the Calculation of Wealth

1. J. E. Stiglitz, A. Sen, and J. P. Fitoussi, *Report by the Commission on the Measurement of Economic Performance and Social Progress*, 2009.

2. D. Méda, *La mystique de la croissance, comment s'en libérer* (Paris: Flammarion, 2013).

3. The economic welfare indicator incorporates the positive and negative elements into aggregate consumption in national accounts. The Index of Sustainable Economic Welfare takes account of deductions from and contributions to the total capital stock that only partially include natural resources. See William D. Nordhaus and James Tobin, "The Measurement of Economic and Social Performance," in *Is Growth Obsolete?* ed., Milton Moss (New York: NBER, 1973), 509–564. The index was subsequently modified, in particular to take better account of natural resources.

4. Edward Barbier and Anil Markandya, *A New Blueprint for a Green Economy* (London: Earthscan, 2012). See, in particular, chap. 2, pp. 18–35.

5. Gene Grossman and Alan Krueger, "Economic Growth and the Environment," *Quarterly Journal of Economics* 110, no. 2 (1995): 353–377.

6. Bernard Chevassus-au-Louis (2013) points out the difficulty of comparison methods, hence the wide range he gives for assessing this rate of change: from a few dozen up to 10,000!

7. For a full description of the initial methodology, see William Rees and Mathis Wackernagel, *Our Ecological Footprint: Reducing Human Impact on the Earth* (British Columbia, Canada: New Society Publishers, 1996). For the use of this indicator, see the Living Planet Report, regularly produced by the WWF and published on its website.

5. "Natural Capital" Revisited

1. The rare earths are a group of metals, including lanthanum, yttrium, and neodymium. Contrary to what their name suggests, they are not "rare," but their concentrations vary and they are very difficult to extract and exploit. Without them, there would be no flat screens, solar panels, smartphones, wind turbines, or catalytic converters.

2. See, in particular, Robert Solow, "Sustainability: An Economist's Perspective," in *Economics of the Environment: Selected Readings*, 3rd ed., ed. Robert and Nancy Dorfmann (New York: W. W. Norton, 1993). This idea of greater or lesser substitutability between factors of production has resulted in a large economic literature on the ideas of sustainability, in both the strong sense and the weak sense.

3. Bernard Chevassus-au-Louis, *La Biodiversité, c'est maintenant* (La Tour d'Aigues: L'Aube, 2013), 77.

4. Two accessible and non-scientific overviews of this work have been published: J. Rockström et al., "Planetary Boundaries: Exploring the Safe Operating Space for Humanity," *Ecology & Society* 14, no. 2 (2009): 32, and J. Rockström et al., "A Safe Operating Space for Humanity," *Nature* 461 (2009).

5. The best known threshold is a maximum concentration of 450 parts per million (ppm) of greenhouse gas emissions in the atmosphere, if warming is to be kept to no more than 2°C.

6. Hotelling: Beyond the Wall of Scarcity

1. ENERGY.WP.1_77, United Nation International Framework Classification for Reserves/Resources—Solid Fuel and Mineral Commodities, 1997.

2. Proven oil reserves rose from 998.4 billion barrels in 1992 to 1,668.9 billion barrels in 2012 (BP, Statistical Review of World Energy, 2013).

3. Harold Hotelling, "The Economics of Exhaustible Resources," *Journal of Political Economy* 39 (1931): 137–175.

4. In other words, for the individual the possibilities are strictly equivalent.

5. Margaret E. Slade, "Trends in Natural-Resource Commodity Prices: An Analysis of the Time Domain," *Journal of Environmental Economics and Management* 9 (1982): 122–137.

6. When we consider the case of renewable resources, the same logic applies. Indeed, the fundamental rule for managing a renewable resource is based on a tradeoff idea similar to the Hotelling rule: the stock of a renewable resource can be likened to a capital stock for which the operator seeks, at equilibrium, a return identical to those from the other assets, namely the interest rate, the basic difference being that it is possible to increase, decrease, or keep constant the stock in question—in other words, the possibility of exploiting the resource below, above, or exactly at its renewal rate.

7. P. Dasgupta and G. Heal, "The Optimal Depletion of Exhaustible Resources," *Review of Economic Studies* 41 (1974): 1–28; R. Solow, "Intergenerational Equity and Exhaustible Resources," *Review of Economic Studies* 41 (1974): 29–45; J. Stiglitz, "Growth with Exhaustible Natural Resources: Efficient and Optimal Growth Paths," *Review of Economic Studies* 41 (1974): 123–137.

8. The term "externality" appeared in the literature in 1957 with the paper by F. Bator, "The Simple Analytics of Welfare Maximization," *The American Economic Review* 47 (1957): 22–57. The Nobel economics prize winner J. Meade defines external effects as changes in an agent's utility generated by the actions of another agent without giving rise to market transactions in his paper "External Economies and Diseconomies in a Competitive Situation," *Economic Journal* 62 (1952): 54–67.

9. Some authors also add primary services such as photosynthesis, the water cycle, and the cycle of nutrients essential to life as well as cultural services (creativity, inspiration, educational values, recreation activities) to this list.

10. Millennium Ecosystem Assessment (MEA), *Ecosystems and Human Well-Being* (Washington, D.C.: Island Press, 2003).

11. See the collective work *Climate Economics in Progress 2013*, Climate Economics Chair, Paris.

12. Carbon dioxide (CO_2) accounts for just under 70 percent of all anthropogenic global emissions. Other greenhouse gas emissions from human activities are expressed as a CO_2 equivalent on the basis of their global warming potential over a hundred years, hence the term "CO_2 equivalent."

7. Nature Has No Price: How Then Is the Cost of Its Degradation to Be Measured?

1. Cited in "Quel prix pour la nature?," the report by Terraeco on www .terraeco.net/a3421.html.

2. These two theorems express the fundamental results of the theory of general equilibrium formalized by the Nobel economics prize winners Kenneth Arrow and Gerard Debreu. K. J. Arrow and G. Debreu, "Existence of an Equilibrium for a Competitive Economy," *Econometrica* 22 (1954): 265–290.

3. A situation in which a single individual possesses almost all the wealth of the economy may well be Pareto optimal.

4. Garrett Hardin, "The Tragedy of the Commons," *Science* 162, no. 3859 (December 13, 1968): 1243–1248.

5. A. Leopold, *A Sand County Almanac* (New York: Oxford University Press, 1949). This posthumous work strongly influenced the development of today's environmental ethics and the whole movement for the protection of natural areas. Aldo Leopold is considered one of the founding fathers of the management of environmental protection in the United States.

6. B. Weisbrod, "Collective-Consumption Services of Individual-Consumption Goods," *Quarterly Journal of Economics* 78 (1964): 471–477.

7. K. J. Arrow and A. C. Fisher, "Environmental Preservation, Uncertainty and Irreversibility," *Quarterly Journal of Economics* 88 (1974): 312–319; C. Henry, "Investment Decisions Under Uncertainty: The Irreversibility Effect," *American Economic Review* 64 (1974): 1006–1012.

8. Beyond Hotelling: Natural Capital as a Factor Required for Growth

1. Formally, the production level *Y* is defined from the combination of capital *K* and of labor *L* in accordance with *F(.)*: *Y = F(K, L)*. If a global factor *A(.)* is taken into account, the function becomes *Y = A(.)F(K, L)*.

2. Robert M. Solow, "A Contribution to the Theory of Economic Growth," *Quarterly Journal of Economics* 70, no. 1 (1956): 65–94; Robert E. Lucas Jr., "On the Mechanics of Economic Development," *Journal of Monetary Economics* 22 (1988): 3–42; Sergio Rebelo, "Long-Run Policy Analysis and Long-Run Growth," *Journal of Political Economy* 99, no. 3 (1991): 500–521; Paul Romer, "Endogenous Technological

Change," *Journal of Political Economy* 98, no. 5 (1990): S71–S102; Robert J. Barro, "Determinants of Economic Growth: A Cross-Country Empirical Study" (working paper, National Bureau of Economic Research, 1996, no. 5698).

3. A Cobb-Douglas production function allows the effects of each factor of production to be analyzed according to its relative weight in wealth creation, $Y = A\ K^a H^b$, with a and b being the weight of physical capital and human capital in production and income distribution respectively. It is generally considered that a represents approximately one-third and b two-thirds.

4. D. Acemoğlu, *Introduction to Modern Economic Growth* (Princeton: Princeton University Press, 2009); P. Dasgupta and S. Niggol Seo, "Natural Capital and Economic Growth," *The Encyclopedia of Earth*, ed. Cutler J. Cleveland (Environmental Information Coalition, National Council for Science and the Environment); first published in *The Encyclopedia of Earth*, August 21, 2008; reissued January 30, 2013.

5. Brian R. Copeland and Scott M. Taylor, "North-South Trade and the Environment," *Quarterly Journal of Economics* 109 (1994): 755–787; Nancy L. Stokey, "Are There Limits to Growth?," *International Economic Review* 39 (1998): 1–31; Pierre-André Jouvet, Philippe Michel, and Gilles Rotillon, "Optimal Growth with Pollution: How to Use Pollution Permits?," *Journal of Economic Dynamics and Control* 29 (2005): 1597–1609.

6. Many studies consider a three-factor production function but generally the third factor is either land (in the agricultural sense), an exhaustible resource, or energy. We propose to expand the scope of natural capital, thus including all the environmental damage associated with human production activities.

7. T. Bréchet and P. A. Jouvet, "Why Environmental Management May Yield No-Regret Pollution Abatement Options," *Ecological Economics* 68 (2009): 1770–1777.

9. Water, the Shepherd, and the Owner: A Choice of Green Growth Models

1. See G. Rotillon, *Faut-il croire au développement durable?* (Paris: L'Harmattan, 2008).

2. The latter represents a form of physical capital. It may seem shocking, here as elsewhere, to consider these animals as physical capital. In fact, sheep are simply the owner's input to the production of fleece. We can nevertheless point out that animals do not seem to be considered other than in the

context of rearing in batteries. With regard to this topic, see the defenders of animal rights, such as "antispeciesist" collectives.

3. See, in particular, Pierre-André Jouvet and Boris Solier, "An Overview of CO_2 Cost Pass-Through to Electricity Prices in Europe," *Energy Policy* 61 (2013): 1370–1376.

4. Within the limit of the emissions cap.

10. How Much Is Your Genome Worth?

1. Published in 1962, Carson's book describes a world in which birds are disappearing because of the widespread use of chemical pesticides. The gradual ban in the United States on agricultural use of DDT beginning in the 1970s was attributable to the impact that the book had on public opinion.

2. See www.teebweb.org as well as Commissariat général au développement durable, *L'Environnement en France*, 2010, for the case of France.

3. The study by Robert Costanza ("The Value of the World's Ecosystem Services and Natural Capital," *Nature* 387 [1997]: 253–260) estimates the services provided to humanity by biodiversity to be worth about $33 trillion. But this estimate has been very much disputed with regard to what it is based on.

4. In fact, we have indicators of biodiversity through the counting of different species. We measure biodiversity loss rather than the benefits it gives us. We also take specific species as a proxy for biodiversity, as in the paper by Chris van Swaay and M. Warren, "Prime Butterfly Areas of Europe: An Initial Selection of Priority Sites for Conservation," *Journal of Insect Conservation* 10 (2006): 5–11.

5. The report for the Centre d'analyse stratégique by Bernard Chevassus-au-Louis, *L'Approche économique de la biodiversité et des services liés aux écosystèmes* (2009), makes the difficulty, or even the impossibility, of a global measure of biodiversity very clear.

6. With thanks to Professor Michel Bera of the "Statistical modeling of risk" Chair of CNAM.

7. Deoxyribonucleic acid (DNA) formed by the sequence of 3 billion base pairs (nucleotides), defining 25,000 genes.

8. See the Foundation's report on research on biodiversity: www .fondationbiodiversite.fr/images/stories/telechargement/rapport_ valeurs_.pdf.

9. In reference to the famous British economist Alfred Marshall, who defined it in his book *Principles of Economics* (London: Macmillan, 1890).

10. K.-G. Mäler, "A Method for Estimating Social Benefits from Pollution Control," *Swedish Journal of Economics* 73 (1971): 121–133.

11. British economist and Nobel economics prize winner in 1972 with Kenneth Arrow, J. R. Hicks, "The Four Consumer's Surpluses," *The Review of Economic Studies* 1 (1943): 31–41.

12. The "fully" aspect is purely theoretical, and it is generally a matter of approximations allowing "acceptable" levels to be measured. This logic is followed in measuring indemnities of all kinds: evictions, compensation for pollution, etc. It is very rare that the parties involved are fully satisfied with the agreement. But as we have said, refusing all evaluations is tantamount to having a zero value, and it then becomes almost impossible to take environmental dimensions into account in decision making.

13. See J. Horowitz and S. McConnell, "A Review on WTA/WTP Studies," *Journal of Environmental Economics and Management* 44 (2002): 426–447.

14. R. Mitchell and R. Carson, *Using Surveys to Value Public Goods: The Contingent Valuation Method* (Washington, D.C.: Johns Hopkins University Press for Resources for the Future, 1989).

15. F. Kong, H. Yin, and N. Nakagoshi, "Using GIS and Landscape Metrics in the Hedonic Price Modeling of the Amenity Value of Urban Green Space: A Case Study in Jinan City, China," *Landscape and Urban Planning* 79 (2007): 240–252. This study, without giving precise figures, underlines the growing importance attached by the Chinese population to having green areas nearby.

16. P. Bayer, N. Keohane, and C. Timmins, "Migration and Hedonic Valuation: The Case of Air Quality," *Journal of Environmental Economics and Management* 58 (2009): 1–14.

17. A. Bengochea Morancho, "A Hedonic Valuation of Urban Green Areas," *Landscape and Urban Planning* 66 (2003): 35–41.

18. See C. Kolstad, *Environmental Economics*, 2nd ed. (New York: Oxford University Press, 2010), for details on methods, or P. Bontems and G. Rotillon, *L'Économie de l'environnement*, 3rd ed. (Paris: La Découverte, 2007), for an overview.

19. See, for example, W. D. Shaw and P. Jakus, "Travel Cost Models of the Demand for Rock Climbing," *Agricultural and Resource Economics Review* (1996), or K. G. Willis and G. D. Garrod, "An Individual Travel Cost Method of Evaluating Forest Recreation," *Journal of Agricultural Economics* 25 (2008): 133–142.

20. See G. Dionne and M. Lebaux, "La valeur statistique d'une vie humaine," *CIRRELT* 48 (2010): 991–1011.

21. In the literature, usually referred to as methodological or econometric "bias."

22. In order to decide between the oil company and the victims with regard to damage assessment, the National Oceanic and Atmospheric Administration (NOAA), in charge of defining the legal methods in the assessment of damages, appealed to a panel of experts chaired by two Nobel prize winners: K. Arrow and R. Solow. The conclusions of the panel gave legitimacy to contingent valuation for legal matters, provided that the protocol of application specified in the report is respected.

23. L. Carr and R. Mendelsohn, "Valuing Coral Reefs: A Travel Cost Analysis of the Great Barrier Reef," *Ambio* 32 (2003): 353–357.

24. See Marit E. Kragt, Peter C. Roebeling, and Arjan Ruijs, "Effects of Great Barrier Reef Degradation on Recreational Reef-Trip Demand: A Contingent Behaviour Approach," *Australian Journal of Agricultural and Resource Economics* 48 (2009): 419–443.

25. *L'Environnement en France*, Commissariat général au développement durable, 2010.

11. The Enhancement of Biodiversity: Managing Access, Pricing Usage

1. H. S. Gordon, "The Economic Theory of a Common Property Resource: The Fishery," *Journal of Political Economy* 62 (1954): 124–142; M. B. Schaefer, "Some Aspects of the Dynamics of Populations Important to the Management of the Commercial Marine Fisheries," *Bulletin of the Inter-American Tropical Tuna Commission* 1 (1954): 25–56; M. B. Schaefer, "Some Considerations of Population Dynamics and Economics in Relation to the Management of Marine Fisheries," *Journal of the Fisheries Research Board of Canada* 14 (1957): 669–681.

2. A. D. Scott, "The Fishery: The Objectives of Sole Ownership," *Journal of Political Economy* 63 (1955): 116–124; C. W. Clark, "The Economics of Overexploitation," *Science* 181 (1973): 630–634.

3. In the Pacific in the early 1990s, halibut fishing was permitted for only three days per year.

4. On the theoretical aspects, see P.-A. Jouvet and G. Rotillon, "Ressources renouvelables et quotas transférables dans un modèle à générations imbriquées," *Recherches économiques de Louvain* 1 (2005): 117–130. For the respective merits of different forms of regulation, see L.-P. Mahe and C. Ropars, "L'exploitation régulée d'une ressource renouvelable: inefficacité d'un rationnement factoriel et efficacité des quotas individuels transférables," *Économie et prévision* 148 (2001): 141–156.

5. Ecosystem Marketplace, *State of Biodiversity Markets* (Washington, D.C.: Forest Trends, 2012).

6. "Fiscalité et artificialisation des sols," adopted on March 28, 2013, by the Comité pour la Fiscalité Ecologique.

7. *Agriculture et biodiversité: Valoriser les synergies* (Versailles: Quae, 2012).

8. We thank Jean-René Brunetière and George Feterman for their input on this sensitive topic.

9. Bernard Chevassus-au-Louis, *La Biodiversité, c'est maintenant* (La Tour d'Aigues: L'Aube, 2013).

10. Bernard Chevassus-au-Louis (under the direction of), *Approche économique de la biodiversité et des services liés aux écosystèmes* (Paris: Centre d'analyse stratégique/La Documentation française, 2009).

11. According to the Regional Agency for Nature and Biodiversity, the Ile-de-France Bee Observatory shows that over the last three years the bee population is doing better in the center of Paris than in its periphery. www .natureparif.fr.

12. The diversity of characteristics of individuals and diversity of macro-ecosystems correspond respectively to phenotypic diversity and to biomes.

12. Climate Change: The Challenges of Carbon Pricing

1. We here drew directly on A. Delbosc and C. de Perthuis, *Et si le changement climatique nous aidait à sortir de la crise économique* (Paris: Le Cavalier bleu, 2012). The reader will find a more comprehensive account of these questions in C. de Perthuis, *Et pour quelques degrés de plus*, 2nd ed. (Montreuil-sous-Bois: Pearson, 2011).

2. The extent of the range results mainly from the way in which uncertainty and the risks associated with extreme scenarios are treated. See N. Stern et al., *The Economics of Climate Change* (Cambridge: Cambridge University Press, 2007) (the complete report is available on the U.K. Treasury website).

3. Ostrom's approach reminds us, however, that it is not necessary to wait for an international climate agreement or the application of a single instrument. See, in particular, "Polycentric Systems: Multilevel Governance Involving a Diversity of Organisation," in *Global Environmental Commons*, ed. E. Brousseau (Oxford: Oxford University Press, 2012), 105–125.

4. The action of CFC gases on the ozone layer was highlighted by American scientists in the mid-1970s. Initially denied vehemently by the manufacturers producing these substances, it gave rise to the search for alternative

products, which the American company Dupont was the first to discover. The finding of alternatives facilitated the adoption of the Montreal Protocol (signed in 1987 under the auspices of the United Nations), which successfully proscribed the use of these substances in industry worldwide. Successful though it was, the Montreal Protocol is not really applicable to climate change because it involved regulating a very limited number of industrial sources around the world, whereas the sources of greenhouse gas emissions take many forms.

5. *Technical Update of the Social Cost of Carbon for Regulatory Impact Analysis Under Executive Order 12866,* Interagency Working Group on Social Cost of Carbon, United States Government, May 2013.

6. A. Quinet (under the direction of), *La Valeur tutélaire du carbone* (Paris: La Documentation Française, March 2009).

7. The pioneering paper by Martin Weitzman, "Prices vs. Quantities" (*The Review of Economic Studies* 41, no. 4 [October 1974]: 477–491), shows that the costs of error are higher in the case where the slope of the marginal cost of abatement is greater than that of marginal benefits, a conclusion that favors the use of tax with regard to climate change.

8. The list of these countries and the reduction commitments they made are detailed in Annex B of the Protocol.

9. The Commission tabled such a "backloading" proposal in 2011, but did not manage to get it passed by the European Parliament until July 2013. Such inertia in decision making is clearly not conducive to seriously regulating a market. See Christian de Perthuis and Raphaël Trotignon, "Governance of CO_2 Markets: Lessons from the European Emission Trading Scheme," *Energy Policy* 75 (Autumn 2013): 100–106.

10. A fairly extensive literature has developed around this issue: L. H. Goulder, "Environmental Taxation and the 'Double Dividend': A Reader's Guide" (working paper, NBER, 1994, no. 4896), and L. H. Goulder, "Effects of Carbon Taxes in an Economy with Prior Tax Distortions: An Intertemporal General Equilibrium Analysis," *Journal of Environmental Economics and Management* 29 (1995): 271–297.

11. Introduced on January 1, 2008, the system taxes purchases of new high-emitting vehicles ("malus"). With the proceeds of the tax, the government funds a rebate ("bonus") on non- or low-CO_2-emitting vehicles. The mechanism is supposed to balance in budgetary terms, but this was not the case during the first years of its implementation due to substantial deferment of purchases until such a time as less polluting vehicles were available.

12. Mireille Chiroleu-Assouline, Le Double Dividende. Les approches théoriques, PhD diss., Université Paris-I, EUREQua and ERASME, 2008.

13. The problem is much more complex for emissions and storage of greenhouse gas emissions from agriculture and forestry.

13. International Climate Negotiations

1. An economic agent adopting free-riding behavior benefits from a (usually public) good financed by other users without paying his share, for example by using public transport without buying a ticket. The concept was formalized and generalized by Mancur Olson in *The Logic of Collective Action* (Cambridge, MA: Harvard University Press, 1965).

2. See the excellent account by William Nordhaus, *The Climate Casino: Risk, Uncertainty and Economics for a Warming World* (New Haven and London: Yale University Press, 2013).

3. The resolution proposed by Senators Byrd and Hagel (a Republican and a Democrat) was passed unanimously by the Senate 95–0 on July 25, 1997, a few months before the climate conference where the Kyoto Protocol was signed. This resolution in fact sought to prevent the adherence of the United States to any climate agreement, so constraining were the conditions for ratification laid down by the Senators. See 105th Congress, 1st Session, Report N°105–54, GPO.

4. The figures presented in the IPCC Fifth Assessment Report regarding the six greenhouse gases covered by climate change agreements are unequivocal: their average annual growth rate rose from 1.3 percent between 1970 and 2000 to 2.2 percent between 2000 and 2010. See IPCC, WGIII AR5, April 2014. Statistics on CO_2 emissions from energy sources compiled by the International Energy Agency and the Oak Ridge National Laboratory, United States Department of Energy, lead to the same conclusions. See Global CO2 Emissions from Fossil-Fuel Burning, Cement Manufacture, and Gas Flaring: 1751–2010, Tom Boden and Bob Andres, CDIAC, Oak Ridge National Laboratory, U.S. Department of Energy, July 2014.

5. The substitute strategies view implies that each player thinks he can benefit from the strategies of others without effort, in contrast to an alternative view whereby each player believes his strategy will prove more effective if it conforms to those of other players. These are the two outlooks underpinning free-rider behavior. T. Sandler, *Global Collective Action* (Cambridge: Cambridge University Press, 2004).

6. The IPCC was established in 1987 under the auspices of two UN agencies: the World Meteorological Organization and the United Nations

Environment Programme. The IPCC is not an additional research center but a network linking up scientists around the world. The First Assessment Report published in 1990 played a key role in the signing of the 1992 Convention on Climate Change. The final conclusions of the Fifth IPCC Report were made public in October 2014 and provide support for the COP 21, due to be held in Paris in December 2015. In addition to its assessment function, the IPCC plays an important role in setting standards for the calculation and accounting of greenhouse gas emissions.

7. The third edition of Global Climate Change Impacts in the United States was published in May 2014. The report, placed under the authority of the National Science and Technology Council, was supervised by more than three hundred American scientists, many of whom also participate in the work of the IPCC. It is intended for Congress and the president of the United States, with numerous illustrations and summaries designed to facilitate understanding by non-climatologist policy makers and elected officials.

8. Founded in 1956 in the United Kingdom by Samuel Shenton, the Flat Earth Society had several thousand members in the 1960s. According to its promoters, who continue to maintain a website, it still numbered a few hundred in 2014.

9. The Yale economist William Nordhaus has been a leading pioneer in climate economics. Any reader wishing to get an overview on these matters should consult William Nordhaus, *The Climate Casino: Risk, Uncertainty, and Economics for a Warming World* (New Haven and London: Yale University Press, 2013).

10. The reader will find highly relevant information on this question in J. G. Shepherd, "Geoengineering the Climate: Science, Governance and Uncertainty," *The Royal Society* (September 2009).

11. Graciela Chichilnisky and Geoffrey Heal, *Environmental Markets, Equity and Efficiency* (New York and Chichester, UK: Columbia University Press, 1998).

12. Thomas Courchene and John Allen, "Climate Change: The Case for a Carbon Tariff/Tax," *March Policy Options* 59 (2008): 59–64.

13. On this point, it is worth looking at Eric Brousseau, Tom Dedeurwaerdere, Pierre-André Jouvet, and Marc Willinger, *Global Environmental Commons: Analytical and Political Challenges in Building Governance Mechanisms* (New York: Oxford University Press, 2012).

14. Wen Wang, "Overview of Climate Change Policies and Prospects for Carbon Markets in China," *Cahiers of the Climate Economics Chair*, Informations & Débats series, no. 18 (July 2012).

15. Richard B. Stewart and Jonathan B. Wiener, *Reconstructing Climate Policy: Beyond Kyoto* (Washington, D.C.: The American Enterprises Institute, 2003).

14. The "Energy Transition": Not Enough or Too Much Oil and Gas?

1. We here make reference to the title of the book by Henri Prévot, *Trop de pétrole! Énergie fossile et réchauffement climatique* (Paris: Seuil, 2007).

2. Robert Hefner, *The Grand Energy Transition – GET: The Rise of Energy Gases, Sustainable Life and Growth, and the Next Great Economic Expansion* (Edison, NJ: Wiley, 2009).

3. This legislative package running to several hundred pages specifies how the "three times twenty" objectives for 2020—reducing greenhouse gas emissions by 20 percent compared to 1990, raising the share of renewable energy in consumption to 20 percent, and increasing energy efficiency by 20 percent—are divided up among the twenty-seven Member States of the European Union.

4. Vaclav Smil, *Energy Transitions: History, Requirements, Prospects* (Westport, CT: Praeger, 2010).

5. Jeremy Rifkin, *The Third Industrial Revolution: How Lateral Power Is Transforming Energy, the Economy, and the World* (Basingstoke, UK: Palgrave Macmillan, 2011).

6. See R. G. Gordon, "Does the New Economy Measure Up to the Great Inventions of the Past?," *Journal of Economic Perspectives*, 14 no. 4 (Fall 2000): 49–74, and "Is US Economic Growth Over? Faltering Innovation Confronts the Six Headwinds," *CEPR, Policy Insight*, 63 (September 2012): 1–13.

7. R. Fouquet and P. Pearson, "Past and Prospective Energy Transitions: Insights from History," *Economics of Energy and Environmental Policy* 1 (2012): 83–100. According to these authors, the price of lighting in the United Kingdom fell by a factor of three thousand over two centuries.

8. William Stanley Jevons, *The Coal Question: An Inquiry Concerning the Progress of the Nation and the Probable Exhaustion of Our Coal Mines* (London: Macmillan, 1865), 140.

9. Smil, *Energy Transitions*, 107.

10. The corresponding scenarios in IPCC studies indicate average temperature increases well above two degrees, usually seen as the threshold that should not be exceeded.

11. We have developed this point in various papers. See Pierre-André Jouvet and Christian de Perthuis, "La croissance verte: de l'intention à la mise en œuvre," *Cahiers of the Climate Economics Chair*, Informations & Débats series, no. 15 (June 2012).

12. The more specialized literature refers to CCS (carbon capture and storage) techniques.

13. Given the warming potential of methane, twenty-five times that of carbon dioxide, an overall leakage rate of 3 percent would be enough to counteract the advantage of gas over coal.

14. With this method the gas has to be liquefied by compression when it is loaded into LNG tankers and then regasified on arrival. Both methods (especially the first) consume a large amount of fossil fuel.

15. See the paper putting forward this argument: R. Howarth, R. Santoro, and A. Ingraffea, "Venting and Leaking of Methane from Shale Gas Development: Response to Cathles et al.," *Climatic Change* 113 (2012): 537–549. The MIT experts take a more nuanced position, probably closer to reality: Francis O'Sullivan and Sergey Paltsev, "Shale Gas Production: Potential versus Actual Greenhouse Gas Emissions," *Environmental Research Letters* 7 (December 2012): 1–6.

16. For gas, the standard term is the British Thermal Unit (BTU), used for trading in the New York market.

17. Wyoming has become the top coal-producing state in the United States, overtaking the traditional deposits of the Appalachians. Its export capacity to Asia is limited by transport bottlenecks (rail links and terminals). Between 2008 and 2012, net exports of coal from the United States increased by more than sixty million tons, accounting for a quarter of the growth in supply on the world market, behind Indonesia but ahead of Australia and Russia.

15. The Inescapable Question of the Price of Energy

1. The hypothesis of a rise in the price of wood linked to forest clearing in the United Kingdom during the eighteenth century was put forward by John Neff in his book *The Rise of the British Coal Industry* (London: Routledge, 1932). This hypothesis was subsequently disputed by a new generation of historians, as shown by Roger Fouquet in *Heat, Power and Light: Revolutions in Energy Services* (Cheltenham, UK: Edward Elgar, 2008).

2. On average, a barrel of oil contains the equivalent of 0.4 tons of CO_2.

3. At this rate, the price of a liter of top-grade gasoline (in France) would rise from €1.7 to €2.15.

4. The term "load factor" is also used to describe the relationship between the actual and the theoretical maximum output. For example, if the load factor of a gas-fired power plant for base use is 85 percent, that of a wind turbine is 20 percent. At the same power, more than four wind turbines are required to produce the same amount of energy as the gas-fired plant.

5. Ernst von Weizsäcker et al., *Factor Five: Transforming the Global Economy through 80% Improvements in Resource Productivity* (London: The Natural Earth Project/Earthscan, 2009).

6. R. Fouquet and P. Pearson, "Past and Prospective Energy Transitions: Insights from History," *Economics of Energy and Environmental Policy* 1 (2012): 83–100.

7. The invention of the light bulb by the Scotsman John Lindsay dates from 1835 and was therefore well before the discovery of the filament bulb process, patented by Edison in 1878, which gave rise to the growth of what is now known as the incandescent bulb.

8. In more technical terms, the income elasticity of demand for lighting rose above two between 1870 and 1930 and remained higher than one until 1950. Price elasticity, for its part, was higher than one until 1900.

9. This is the thesis put forward by Brookes and Khazzoom: L. Brookes, "Energy Efficiency and Economic Fallacies," *Energy Policy* 18, no. 2 (1990): 199–210; D. J. Khazzoom, "Economic Implications of Mandated Efficiency Standards for Household Appliances," *Energy Journal* 1 (1980): 21–40. The Brookes-Khazzoom postulate was subsequently challenged on the basis of numerous empirical observations by authors such as Grubb and Lovins.

10. Emissions associated with these sources are highly dependent on the quality of the fuels concerned: lignite coal and heavy oils, for example, emit significantly more than conventional coal and light oils for the same amount of energy provided.

11. In April 2013, due to low CO_2 prices and a decrease in the price of European coal following extensive coal exports from the United States as a result of the development of shale gas, GDF Suez decided to "mothball" two combined cycle gas-fired plants for the summer and another for an indefinite period.

12. In California, the reserve price, below which the market price may not fall, was set at $10 per ton of CO_2 in 2013 and will increase annually by 5 percentage points above the retail price index up to 2020. In France, the 2014 Finance Act provides for the introduction of a carbon tax of €7 per ton of CO_2, rising to €22 in 2016.

13. There are two kinds of risk from CO_2 leaks: those that involve the gradual and imperceptible release of CO_2 into the atmosphere and are harmless to the local population; and concentrated releases, in which health risks may arise. In 1986, the sudden release of a huge pocket of CO_2 from Lake

Nyos, Cameroon, resulted in the death 1,700 local residents. CO_2-related accidents are, however, extremely rare compared to those resulting from the inhalation of carbon monoxide (CO).

14. Gabriela Simonet et al., "Forest Carbon: Tackling Externalities," *Cahiers of the Climate Economics Chair*, Informations & Débats series, no. 17 (July 2012).

16. Nuclear Energy: A Rising-Cost Technology

1. Having decided to extend the life of Germany's nuclear power plants at the beginning of her term, in March 2011 Angela Merkel made the decision to close down the eight oldest reactors and to review the operational lifetimes initially planned for the rest. If no new nuclear power stations are built and if the lifetimes of existing plants are not extended, Germany will have completely phased out nuclear power by 2022.

2. For example, according to Eurostat, for a typical consumer (with annual consumption of between 2.5 and 5 MWh), the price of residential electricity in the first half of 2012 was less than 0.15 euro per kWh in France against more than 0.25 euro per kWh in Germany.

3. Audit, La Cour des Comptes, *Les Coûts de la filière électronucléaire* (Paris: Public Report, January 2012).

4. Because electricity is not easily stored, its production must constantly be adjusted to demand, thereby causing significant fluctuations in its price in wholesale markets. When people return home in the early evening, additional supplies are generated to meet peak demand (typically using gas or coal-fired plants and hydroelectric dams) and the market price per kilowatt hour rises sharply.

5. François Lévêque, *Nucléaire On/Off, Analyse économique d'un pari* (Paris: Dunod, 2013).

6. "The Non-Nuclear Future" (Presentation at the Salzburg Conference, May 1977).

17. Growth-Generating Innovations

1. The "Kinetoscope" was invented in 1888 by Edison and a patent was filed for the United States. Commercial operation of the first cinemas started a few years later. In 1902, a reel of a film made by Georges Méliès,

A Trip to the Moon, was stolen in London, shipped to the United States, and reproduced in multiple copies. Despite a number of lawsuits in New York, Georges Méliès never recovered from this competition by Edison, and his studios went bankrupt.

2. We might also mention food. In meeting our nutritional requirements, food production uses a considerable land area that is therefore no longer available for the production of biomass energy.

3. With the notable exception of major hydro projects. With an installed capacity of 22,500 megawatts, the Three Gorges Dam in China is the most concentrated source of electricity in the world, followed by Itaibu and Giri in Latin America, at over 10,000 megawatts apiece. The world's largest power stations produce no more than 6,000 megawatts, and the Flamanville EPR is programmed at 1,650 megawatts.

4. In this case, the additional cost of photovoltaic production is largely offset by being able to dispense with the work needed for connection to the network.

18. Planning or the Market: What Are the Catalysts?

1. American economist Milton Friedman, winner of the 1976 Nobel Prize in economics, spent his entire career at the University of Chicago. He is considered one of the most influential free market economists of the twentieth century.

2. This example is taken from the "Economics of Natural Resources" course taught by Gilles Rotillon and Pierre-André Jouvet.

3. Ronald Coase, "The Problem of Social Cost," *Journal of Law and Economics* 3 no. 1 (1960): 1–44. This paper is often cited in the literature when seeking theoretical justification for emission allowances markets. Doing so may be justified for voluntary markets that play a very minor role in environmental pricing. But it is quite inappropriate for markets based on decisions by the public authority. Despite his Nobel Prize, Coase is therefore of little help in finding the right way to implement environmental pricing.

4. The committee recommended earmarking 75 percent of the expected revenue from the tax to reduce social security contributions (a double dividend) and 25 percent for flat-rate degressive compensation targeted at vulnerable households.

5. William McDonough and Michael Braungart, *Cradle to Cradle: Remaking the Way We Make Things* (New York: North Point Press, 2002).

6. Ernst von Weizsäcker, "Long-Term Ecological Tax Reform," in *Factor Five: Transforming the Global Economy through 80% Improvements in Resource Productivity*, ed. Ernst von Weizsäcker et al. (London: The Natural Earth Project/Earthscan, 2009), 313–331.

7. T. Jackson, *Prosperity without Growth: Economics for a Finite Planet* (London: Earthscan, 2009), 136–142.

8. William Nordhaus, *The Climate Casino: Risk, Uncertainty and Economics for a Warming World* (New Haven, CT: Yale University Press, 2013), 285.

9. Is there no wiser use for $16 billion of public money (the official estimate for a project initially costed at $10 billion, but for which the final bill is likely to be more than $20 billion) than the construction of ITER (the International Thermonuclear Experimental Reactor)?

10. See the very comprehensive book by E. Brousseau, T. Dedeurwaerdere, P.-A. Jouvet, and M. Willinger, eds., *Global Environmental Commons: Analytical and Political Challenges in Building Governance Mechanisms* (Oxford: Oxford University Press, 2012).

19. European Strategy: Jump Out of the Warm Water!

1. The proportion of public spending and social transfers relatively insensitive to changes in GDP is significantly greater in Europe than in the United States. In the event of recessionary shock, they exert a stabilizing effect in the short term by slowing revenue contraction. At the same time, they may defer the structural adjustments needed to put the economy back into a healthier state.

2. Excise duties are indirect taxes, the amount of which is calculated on the basis of a physical unit (unlike VAT, in which the base is the value of the good concerned). In France, the most important is the TICP (*taxe intérieure de consommation sur les produits*—domestic consumption tax on products), successor to the TIPP (*taxe intérieure sur les produits pétroliers*—domestic tax on oil products), introduced into the tax system by the 1928 Finance Act. The initial purpose was to offset the lower fiscal inflow from salt, itself heir to the famous salt tax in force in France since the Middle Ages. At fourteen billion euros, the TICP was the French State's fourth largest source of revenue in 2012.

3. In this respect, the United States outdid Europe because the American Clean Energy and Security Act, better known as the Waxman-Markey Bill,

which was rejected by the U.S. Senate during Barack Obama's first term, comprised no less than 1,800 pages!

4. Boris Solier, "Une analyse économique et ex-post des effets du prix du carbone sur le secteur électrique" (PhD diss., Paris-Dauphine University, June 2014).

5. Raphaël Trotignon developed the ZEPHYR simulation model for reconstructing past market trends and simulating different future scenarios. See "In Search of the Carbon Price. The European CO_2 Emission Trading Scheme: From Ex Ante and Ex Post Analysis to the Projection in 2020" (PhD diss., Paris-Dauphine University, October 2012).

6. On this second, more technical aspect, see Christian de Perthuis and Raphaël Trotignon, "Governance of CO_2 Markets: Lessons from the European Emission Trading Scheme," *Energy Policy* 75(C) (2013): 100–106.

7. Raphaël Trotignon and Anaïs Delbosc, "Échanges de quotas en période d'essai du marché européen," *Étude climat* 13 (June 2008), Mission Climat de la Caisse des dépôts.

8. See Jean-Marie Chevalier, Michel Derdevet, and Patrice Geoffron, *L'Avenir énergétique: cartes sur table*, Folio actuel (Paris: Gallimard, 2012).

9. Benjamin Dequiedt, "Émissions de gaz à effet de serre et politique agricole commune: quel ticket gagnant?," *Cahiers of the Climate Economics Chair*, Informations & Débats series, no. 12 (January 2012).

Conclusion. Green Capital, Green Capitalism?

1. Thomas Piketty, *Le Capital au XXI siècle* (Paris: Le Seuil, 2013). (English-language edition, *Capital in the Twenty-First Century* [Cambridge: Harvard University Press, 2014].)

2. J. Schumpeter, *Capitalism, Socialism and Democracy* (New York: Harper and Row, 1942).

3. F. Gonand and P. A. Jouvet, "The 'Second Dividend' and the Demographic Structure," *Les Cahiers de la Chaire Economique Working Papers Series*, no. 2014–05 (2014).

Index